Blue Carbon Ecosystems for Sustainable Development

Volume 1

This new, exciting, scientific book series Blue Carbon Ecosystems for Sustainable Development derives from the Springer Tasks for Vegetation Science book series https://www.springer.com/series/6613, especially from those volumes dealing with the Sabkha Ecosystems of the World.

Sabkha ecosystems, under climate change and rising seawater-levels, can gradually convert into salt marshes, mangroves, seagrass beds, and macro-algal reefs, all of which are clearly confined to certain zones above or below chart datum, respectively mean sealevel, depending on location. In coastal hyper-arid regions, sabkhat often form the surface of very thick and old layers of blue-green algal mats. With an increase of precipitation these systems can convert into brackish ponds with reedbeds. All of these hypersaline ecosystems are blue carbon ecosystems, many of them with high productivity rates and of major global importance for carbon sequestration and accumulation of organic matter in the soil.

Furthermore, these blue carbon ecosystems are of tremendous value toward the utilization of high-salinity water, such as seawater, as well as hyper-saline soils, and salt-tolerant biota for the production of biomass, for food, feed, fiber, fuel, other economic purposes, and carbon sequestration.

In order to achieve the 17 United Nations Sustainability Goals, the scientific research into blue carbon ecosystems is an absolutely important element. Moreover, monitoring is needed, as well as the conservation of the remaining blue carbon ecosystems, and the restoration of lost habitat.

The editors of the book series are happy to generate with the volume editors and authors comprehensive and new scientific knowledge on Blue Carbon Ecosystems and their components, with a special view to halophytes and salinity. The book series will provide comprehensive scientific documentation inspiring the way forward on how to utilize saline resources in the best interest of humanity. The next needed steps are to foster academic research and networking, involve land-users, politicians, development banks, innovative farmers, investors, energy-producers, and carbon-off-setters to get involved.

Other than the former Sabkha Ecosystems of the World series, this Blue Carbon Ecosystems for Sustainable Development series will have two major innovations:

1. The geographical realm is open for all global compartments beyond the sub-tropical and tropical regions that were previously covered.
2. The title of this new series clearly highlights the globally important issue of climate change and nature-based solutions of carbon sequestration in support of achieving the United Nations Sustainable Development Goals.

Miguel Clüsener-Godt • Hiroyuki Matsuda
Benno Böer • Ronald A. Loughland
Editors

Blue Carbon Mangrove Ecosystems

A Concept-Based Approach

Springer

Editors
Miguel Clüsener-Godt
UNESCO Chair on Biodiversity
Safeguard for Sustainable Development,
CFE, Centre for Functional Ecology –
Science for People & the Planet,
Department of Life Sciences
University of Coimbra
Coimbra, Portugal

Benno Böer
UNESCO New Delhi Office
in South Asia for Bangladesh,
Bhutan, India, Nepal, the Maldives,
and Sri Lanka
New Delhi, India

Hiroyuki Matsuda
Institute for Multidisciplinary Studies
Yokohama National University
Yokohama, Kanagawa, Japan

Ronald A. Loughland
UNDP
Riyadh, Saudi Arabia

ISSN 2730-5430 ISSN 2730-5449 (electronic)
Blue Carbon Ecosystems for Sustainable Development
ISBN 978-3-031-69552-0 ISBN 978-3-031-69553-7 (eBook)
https://doi.org/10.1007/978-3-031-69553-7

This Springer imprint is published by the registered company Springer Nature Switzerland AG
The registered company address is: Gewerbestrasse 11, 6330 Cham, Switzerland

Foreword

The report entitled "Blue Carbon: The Role of Healthy Oceans in Binding Carbon: A Rapid Assessment" edited by Nellemann C. et al. (2009) highlighted that the world's oceans bind an estimated 55% of all carbon in living organisms. The ocean's blue carbon sinks—particularly, mangroves, marshes, and seagrasses, capture and store most of the carbon buried in marine sediments. This is called "blue carbon". Carbon removed via photosynthesis and stored in plants is called "green carbon". These ecosystems, however, are being degraded and are disappearing at rates 5–10 times faster than rainforests.

Together, by halting the degradation of ecosystems that bind "green" and "blue" carbon, emission reduction is equivalent to 1–2 times that of the entire global transport sector—or at least 25% of the total global carbon emission reductions, with additional benefits for biodiversity, food security, and livelihoods. After this report was published, many organizations and individuals have started to use the term "blue carbon" to emphasize its importance as blue carbon sinks in mitigation of climate change.

The International Society for Mangrove Ecosystems (ISME) was established in 1990 as an international non-governmental organization with its headquarters in Okinawa, Japan. ISME has been working on scientific research, conservation, and restoration of mangrove ecosystems, and has conducted projects in many countries for over 30 years. ISME continues to emphasize on the urgent need for the conservation and sustainable management of existing mangroves, and for the restoration of degraded mangrove ecosystems. Recently, ISME has been supporting mangrove restoration projects aimed at getting carbon credits to accelerate blue carbon mitigation activities.

Besides the role of mangroves as carbon sinks, it should not be forgotten that these ecosystems are also important because of their high primary productivity rates and the habitats they provide for a wide range of fauna and flora. Of equal importance are the goods and services rendered to coastal human communities and their livelihoods. They include timber and non-timber products, fisheries, medicine, recreation, tourism, biological filtration, and coastal protection.

The loss of coastal ecosystems as blue carbon sinks is a global issue that needs to be addressed urgently in view of the serious negative consequences on coastal human communities and living organisms that depend on the sustenance ecosystem services for their livelihoods.

I hope this book series on "Blue Carbon Ecosystems for Sustainable Development" will provide the necessary scientific information on the importance of blue carbon ecosystems and on the immediate actions needed to address global climate change and to save the global environment.

All of us humans have to learn again to live in harmony with nature, a sustainable lifestyle known as *Satoyama* and *Satoumi* in Japan.

Executive Director, International Society for Mangrove Ecosystems (ISME), Okinawa, Japan

Shigeyuki Baba

Preface

It gives me a great pleasure to present to you this new scientific book series "Blue Carbon Ecosystems for Sustainable Development", which all the authors of the Sabkha Ecosystems Book Series publish again with Springer, continuing thus the gathering of extremely important information of coastal ecosystems of the world. Climate change and rising seawater levels will certainly change many of the coastal areas of the world and affect the living conditions of local populations. This change can go along with an increase or decrease of precipitations, which can change considerably the ecology of these systems, having in consequence a heavy impact on coastal zone management, shore protection, carbon sequestration, and accumulation of organic matter in the soil. The production of food, feed, fibre, fuel, and other economic purposes can change irreversibly.

Therefore, it is necessary to study possible impacts by observing ongoing processes and collect scientific information on blue carbon ecosystems. It is an absolutely important need to compile comprehensive scientific information through international conferences and seminars in order to anticipate possible changes and its consequences for the well-being of humanity. The authors will involve the most prominent scientists, land-users, politicians, development bankers, innovative farmers, investors, energy-producers, and carbon-off-setters from all over the world.

This series will be a major scientific contribution for the support of the implementation of the United Nations Sustainable Development Goals.

Major attention, but not exclusively, will be given to UNESCO Sites, i.e. Natural World Heritage Sites, Biosphere Reserves and Geoparks, of which many are located on islands and in coastal areas. As the problems emanating from climate change and sea-level rise are becoming even more important, islands and coastal areas are the first confronted with these phenomena. Biosphere Reserves and other areas implemented by UNESCO are the ideal tools to monitor these changes and to elaborate strategies for mitigating and adapting to these changes, and consequently a direct implementation of the Sustainable Development Goals 2030, the Aichi Targets and the Actions stipulated in the Lima Action Plan for Biosphere Reserves 2016–2025.

In this publication, presentations are outlined that emphasize on how to get coastlines back in balance and to see how we can together overcome the enormous problem for example of plastic pollution.

Paris, France
May 2021

Miguel Clüsener-Godt

Contents

Chapter 1
Advancing Sustainable Development Through Collaboration: Key Takeaways from the International Conference on Blue Carbon Ecosystems

Hiroyuki Matsuda and Miguel Clüsener-Godt

Abstract The International Conference on "Blue Carbon Ecosystems for Sustainable Development with Special Emphasis on Mangrove Ecosystems" was held by Yokohama National University in collaboration with UNESCO and ISME. Blue carbon ecosystems, including mangroves, are critical for carbon sequestration, biodiversity, and supporting high-salinity environments. Key speakers discussed various aspects of these ecosystems, emphasizing their global significance and challenges. Matsuda introduced initiatives in Yokohama, highlighting the evaluation of carbon storage in local marine environments. Kohsaka defined blue carbon ecosystems and their global distribution, noting Indonesia's high potential. Inoue discussed global carbon sequestration in mangroves, while S. Khan focused on blue carbon ecosystems in the Asia-Pacific region and their role in climate change mitigation. Gul addressed the impacts of salinity on mangroves, and Jeong emphasized the need for capacity building to achieve a carbon-neutral society. Abreu discussed conservation and restoration projects in Principe Island's Biosphere Reserve. Zimmer highlighted the importance of quantifying carbon sequestration potential. Hameed and Aragon provided insights on sabkha ecosystems and South-South cooperation, respectively. The conference underscored the need for conservation, restoration, and sustainable management of blue carbon ecosystems. It also stressed the importance of international collaboration, capacity building, and policy support to advance sustainable development and climate change mitigation through these critical ecosystems.

H. Matsuda (✉)
Institute for Multidisciplinary Studies, Yokohama National University, Yokohama, Kanagawa, Japan

M. Clüsener-Godt
UNESCO Chair on Biodiversity Safeguard for Sustainable Development, CFE, Centre for Functional Ecology – Science for People & the Planet, Department of Life Sciences, University of Coimbra, Coimbra, Portugal

M. Clüsener-Godt et al. (eds.), *Blue Carbon Mangrove Ecosystems*, Blue Carbon Ecosystems for Sustainable Development 1,
https://doi.org/10.1007/978-3-031-69553-7_1

Keywords Mangrove conservation · Carbon sequestration · UNESCO biosphere reserves · Coastal ecosystem restoration · Climate change mitigation

The Yokohama National University—UNESCO Chair on "Education in Biosphere Reserves for Sustainable Societies" (EBRoSS) held the International Conference on "Blue Carbon Ecosystems for Sustainable Development with Special Emphasis to Mangrove Ecosystems" on January 11–13, 2023, in collaboration with the UNESCO New Delhi Office and the International Society for Mangrove Ecosystems (ISME).

Saline ecosystems are blue carbon ecosystems, many of them with high productivity rates and of major global importance for carbon sequestration and accumulation of organic matter in the soil. They are of tremendous value toward the utilization of high-salinity water, such as seawater, as well as hyper-saline soils and salt-tolerant biota for the production of biomass, for food, feed, fiber, fuel, other economic purposes, and carbon sequestration. In order to achieve the 17 United Nations Sustainability Goals, scientific research into blue carbon ecosystems is an absolutely important element. Moreover, monitoring is needed, as well as the conservation of the remaining blue carbon ecosystems and the restoration of lost habitat.

We hope to generate new scientific knowledge about the Blue Carbon (BC) ecosystem and its components, with a special focus on saline vegetation and salinity, and especially on mangroves. We also believe that UNESCO's protected areas, World Natural Heritage sites, Biosphere Reserves, and Ramsar Wetlands of International Importance can play important roles in the conservation and engagement of BC ecosystems. This paper summarizes the presentations made at the conference, particularly with regard to the relationship between UNESCO's designated protected areas and the BC Initiatives.

Before the main session, Mr. Nobutoki and Prof. Matsuda introduced the BC project initiatives in Yokohama City. As mentioned later, Yokohama City evaluates the carbon storage capacity of seagrass beds and kelp farms and provides carbon credits for local certification along with other reduction efforts of CO_2 emission. On the other hand, businesses or events that emit CO_2, require carbon offset payments and facilitate intermediation between buyers and sellers of carbon credits.

As the first speaker of this conference, Dr. Kohsaka talked about the "Definition of Blue Carbon Ecosystems," presenting the definition of BC, the global distribution of three blue carbon ecosystems (BCEs), values of BCEs, studies including by himself, and future perspective of BCEs. Indonesia has the highest potential concentrations of BC soil stocks per country (Macreadie et al. 2021). Awareness Level of BCE services is lower in seagrass than mangrove in Busuanga Island, Philippines (Quevedo et al. 2021). In the academic community, blue carbon-themed publications have increased since the conception of the term in 2009 (Quevedo et al. 2023). Finally, Dr. Kohsaka highlighted the role of macroalgae, which are highly productive and have the largest global area of any vegetated coastal ecosystem, in carbon sequestration despite of the fact that there are relatively few cases where they have been included in BC assessments.

Fig. 1.1 Scenery of the meeting. Inoue presented remotely, chaired by Dr. Clüsener-Godt

The second presenter was Dr. Inoue from remote, speaking on "Carbon Sequestration in Mangroves at a Global Level." Carbon sequestration in mangroves occurs through both live biomass and necromass, which includes dead biological materials. Mangrove habitats face various challenges such as high salinity, flooding, anoxia, and strong light, and their aboveground biomass shows a linear decline with increasing latitude, but is higher in regions with higher temperatures and rainfall (Hutchison et al. 2014). Southeast Asia holds nearly half of the global mangrove biomass. Variations exist due to environmental factors like soil nutrients and flooding conditions. Global estimates of carbon storage in mangroves range from <2.6 Pg (Atwood et al. 2017) to 5.0 Pg (Jardine and Siikamäki 2014), with differences stemming from the use of different areal distribution data. Annual global carbon flux in mangroves is estimated to be 26.1 Tg (Breithaupt et al. 2012; Inoue 2019). The stored carbon's status can be influenced by environmental changes, such as deforestation, sea level rise, acidification, and global warming. While substantial data is available for the central Indo-Pacific, information is lacking for regions like South America, West Africa, Central Africa, the Middle East, and Australia, as well as for subsurface biomass and necromass (Fig. 1.1).

Third, Dr. S. Khan, on behalf of one of the organizer and the UNESCO representative person, spoke on "Blue Carbon Ecosystems (BCEs) in the Asia-Pacific Region". It is known that 83% of the global carbon cycle is stored in the ocean, according to the Blue Carbon Initiative (www.thebluecarboninitiative.org), a collaboration between Conservation International, The Intergovernmental Oceanographic Commission (IOC) of UNESCO, and the International Union for Conservation of Nature (IUCN). The BCEs have been found on every continent except Antarctica and cover between 13.8 and 15.2 million hectares (Mha), 2.2 and 40 Mha, and 17.7 and 60 Mha, respectively. Combined, these ecosystems cover approximately 49 Mha (Fig. 1.2).

Coastal BCEs (Blue Carbon Ecosystems) are dynamic ecosystems that serve as crucial habitats, fostering rich biodiversity. They concurrently present actionable

Fig. 1.2 Global distribution of blue carbon ecosystems. (Source: IOC)

avenues for addressing climate change through both mitigation and adaptation strategies. Mangrove ecosystem provides important services including coastal protection, carbon sequestration, improvement of water quality, habitat protection for diverse biota, and fostering blue economy. The greatest net losses were in Southeast Asia (4.8%) and North and Central America and the Caribbean (4.7%) (Spalding et al. 2010). Khan introduced Blue Carbon Perspectives in Asia-Pacific Countries, highlighting their significance in climate change mitigation. Khan emphasized the need for fostering alternative livelihoods, securing financing, enhancing policy frameworks, capacity building, and promoting knowledge sharing to support and advance blue carbon initiatives.

Forth, Prof. Gul talked about "the role of high-salinity-levels for blue carbon ecosystems". Gul pointed out that coherent trends of salinity have been observed and are characterized by global freshening in sub-polar latitudes and a salinification of the shallower parts of the tropical and subtropical oceans. The salinity of coastal waters is strongly influenced by rainfall and river flows. Increasing salinity due to climate change has changed the adaptability of certain mangrove species. Mangrove populations are threatened—by over-exploitation, pollution, and a decline in fluvial discharge into the Indus delta. It is estimated that mangrove cover in the Asia region has decreased by 15% in the past 20 years in part due to a reduction in Indus discharge. Earlier, the Indus delta region had 8 Mangrove species but now only 4 remain. Gul concluded that future mangrove reforestation required mangrove seedlings, selected on the basis of their salinity tolerance.

Fifth, Prof. Jeong talked about "the desirable framework for creating a carbon-neutral society". Jeong's talk is summarized by three points. (1) The concept of Earth's carrying capacity encompasses its ability to provide resources and treat waste, but it has been exceeded globally by 2.5 times. (2) Industrialization, while improving life, has led to resource depletion and pollution, causing a crisis for nature and human existence. Sustainable development has faced limitations and conflicts. (3) Capacity building, through finance, technology, and cooperative networks, is crucial to address these challenges. Industries, civil organizations, media, citizens, and cultural ethos must work together. Collaboration with governments and institutes is vital for joint responses to issues like climate change. Urgent action is required to mitigate the adverse impacts of overexploitation and safeguard Earth's future.

Sixth, Dr. Abreu talked from remote about the Blue Carbon Ecosystem Conservation and Restoration Project at Principe Island's Biosphere Reserve, which aims to protect and restore the valuable coastal habitats that make up blue carbon ecosystems. The project involves several key components: conservation of existing blue carbon ecosystems, restoration of degraded habitats, local community engagement, research and monitoring, capacity building, sustainable financing, and policy support. The ultimate goal of the BCE Conservation and Restoration Project at Principe Island is to mitigate the impacts of climate change by preserving and rehabilitating these critical coastal habitats, while also fostering sustainable development and enhancing the resilience of local communities and ecosystem.

Seventh, Prof. Zimmer talked about "Quantifying the potential of carbon sequestration in Blue Carbon ecosystems." Prof. Zimmer emphasized the significance of accurately and region-specifically quantifying the carbon sequestration potential, as well as fluxes and dynamics of organic matter, as carrier of organic carbon, to comprehend the potential contributions of BCE to climate change mitigation. This data empowers researchers and policymakers to make informed decisions concerning the conservation and re-establishment of BCEs. BCEs offer a potent pathway to achieve net negative carbon emissions, creating co-benefits for both climate action and the blue economy.

Eighth, Dr. Hameed talked on "Sabkha Ecosystems of the World." Sabkha refers to salt-crusted coastal and inland flats with evaporite-saline minerals. Definitions of sabkha have varied, with some restricting it to coastal flats and others differentiating between coastal and inland sabkha (Barth and Böer 2002). Inland sabkha are aeolian in nature and occur in desert plains with episodic flooding from precipitation. Coastal and supralittoral sabkha experience varying degrees of marine tide inundation. Found in arid regions with low precipitation and shallow groundwater levels, sabkha ecosystems support stress-tolerant plants like halophytes, as well as algal and diverse animal life. However, they have been subject to many anthropogenic interventions such as landfilling, top-soiling, dredging, and construction, salt extraction and shrimp farming (Lokier 2013). Sustainable management and conservation efforts are necessary to protect these functional ecosystems. The book series "Sabkha Ecosystems" aims to provide clarity on the various definitions and emphasize their ecological significance.

Ninth, Prof. Aragon talked about "Collaboration with the UNESCO Chair on South-South Cooperation for Sustainable Development in Brazil: Blue Carbon Ecosystems in Coastal Amazonia." Federal University of Para, Brazil, initiated "UNESCO Chair in South-South Cooperation for Sustainable Development." The project fosters South-South cooperation and scientific knowledge to achieve sustainable development goals. It focuses on delivering high-quality information on environmental change, human-environment interactions, and their consequences for well-being. Activities include partnerships in higher education, research, and documentation, with emphasis on the Amazon region and Biosphere Reserves Programme. Facilitating collaboration among renowned researchers and educators from Brazil, Latin America, the Caribbean, and other regions, the project aims to generate concrete results and improve environmental synergy. Its objective is to contribute to informed decision-making and positive impacts on the environment and human communities. Prof. Aragon concluded that The UNESCO Chair in South-South Cooperation for Sustainable Development, can be a catalyzer to facilitate interaction between universities, research institutes and graduate programs in Brazil, in special those located in the Amazon and initiatives of other countries, abording Blue Carbon Ecosystems in the Brazil Amazonian Coast.

On the second day, the location shifted to Hakkeijima, nestled along Tokyo Bay in close proximity to the aquarium. The tenth speaker, Prof. Matsuda, again delivered a presentation titled "Expectations for Japanese Biosphere Reserves and the UNESCO Chair on the Education in Biosphere Reserves for Sustainable Societies (EBRoSS)". While World Heritage is a value-preserving device, a biosphere reserve (BR) is a value-creating institution. BR is a place of mutual learning among local practitioners, researchers, and government officials not only for the conservation of conservation areas, but also for the realization of a sustainable society that utilizes natural resources and is the subject of natural and social science research. The UNESCO Chair on EBRoSS focuses on payments for ecosystem services and studies the interrelationships between urban populations and remote biosphere reserves. Developed countries support environmentally friendly development in developing countries because it is more cost-effective to protect nature in developing countries than to restore nature in developed countries. Assistance to developing countries is not a charity project, but an environmentally friendly project for developed countries.

Eleventh, Dr. Böer spoke on "Floating Mangroves, a new alternative for carbon sequestration". "Floating Mangrove Concept" is an innovative approach for coastal ecosystem restoration and climate change adaptation. It involves creating artificial floating structures that mimic natural mangrove habitats, providing resilience to rising sea levels and erosion. The concept harnesses mangroves' ecological benefits like carbon sequestration and biodiversity support while promoting sustainable economic development through ecotourism and livelihood opportunities. However, the economic feasibility is still under study. We need to determine suitable areas for mangrove production, quantify their carbon sequestration potential, assess their biofuel production capacity and their ability to absorb marine pollution and consider their effectiveness in mitigating tropical storms and tidal waves. In addition, manufacturing and transportation costs should also be evaluated. Research is ongoing investigating how seagrasses, seaweeds and corals can potentially enhance and support mangrove ecosystems.

Twelfth, Dr. Fujimoto talked about "Blue Carbon Systems and Mangroves in Japan." focusing on mangrove ecosystems in Japan, including distribution, carbon production, fine root production, and the impact of sea-level rise. Japanese mangroves have low above-ground biomass due to their small trees, and their fine root production is slow, but belowground carbon storage is comparable to tropical mangroves. Data on belowground carbon storage are gradually accumulating, but specific details on species and site environments must be included. With sea-level rise progressing, surface erosion in low sediment inflow areas is concerning, necessitating observations of its impact on carbon sequestration and surrounding ecosystems.

Thirteenth, Dr. W. R. Khan talked about "Blue Carbon Systems and Mangroves in Malaysia." Malaysia possesses significant coastal blue carbon potential with around 577,558 hectares of mangrove cover, capable of storing approximately 593 million tons of CO2. However, more extensive data on ecosystem distribution is needed for precise estimates. Mangrove management practices vary among states, leading to a lack of sustainable management. The Matang Mangrove Forest Reserve, managed for timber production, faces declining productivity. The Malaysian Action Plan includes initiatives for blue carbon profiling, research, government engagement, and capacity building. Challenges in Malaysia involve a lack of expertise and comprehensive data on blue carbon, necessitating in-house training and integration into existing policies for better management. To address gaps and challenges, a national forum on climate change adaptation and blue carbon policy formulation is considered.

Fourteenth, Dr. Clüsener-Godt talked about "The 30 by 30 challenge and the MAB Programme," which underscores the uncertainty of Earth's species count and the importance of conservation to safeguard biodiversity and ecosystems. It outlines four ways to achieve Impact Equality, including reducing global consumption, depopulation, increasing the efficiency with which the biosphere's supply of goods and services, and investing in nature through conservation and restoration. The Öko-Bauern initiative promotes organic farming for water protection as a good practice of payment for ecosystem services (PES). Restoration efforts are vital for a healthier biosphere. Coastal wetlands like mangroves and seagrass hold significant carbon sequestration potential. The 30 by 30 challenge, established in the Kunming-Montreal Biodiversity Framework at CBD/COP15 in 2022, aims to protect 30% of Earth's land and oceans by 2030 to combat climate change and conserve biodiversity. It also targets reducing global food waste and phasing out harmful subsidies while mobilizing funding for biodiversity conservation. UNESCO's network of Biosphere Reserves and other designated sites plays a crucial role in achieving these goals, necessitating international cooperation and additional site nominations by countries.

Fifteenth, Dr. Böer again talked about "Future perspectives of the MAB Programme in relation to blue carbon ecosystems and mangroves." UNESCO's Preamble emphasizes the importance of constructing defenses of peace in the minds of people. Amid ongoing crises like wars, the COVID-19 pandemic, and climate change, peace is crucial for human survival. The MAB program fosters the relationship between humans and their environment, promoting sustainable economic development and nature conservation. Biosphere Reserves serve as testing grounds for living in harmony with nature, preserving ecosystems and enhancing public awareness. Blue carbon ecosystems like seagrass beds, mangroves, and salt marshes play a role in carbon sequestration, but further research is needed. There are many BRs that include BCEs in each continent. Biosphere Reserves and the MAB Program can raise awareness and support conservation and restoration efforts, scientific research, and sustainable practices, involving communities and private sectors.

Sixteenth, Dr. Hasegawa talked about "Promoting tri-party collaboration through the conservation of mangrove ecosystems of Fonseca Bay in Central America." Biosphere Reserves hold great potential for strengthening transboundary environmental governance, fostering tri-party collaboration, and linking biodiversity and climate issues through the Nexus Approach. The Fonseca Bay Mangrove Conservation Project in Central America exemplifies how Biosphere Reserves can promote collaboration among nations to conserve crucial ecosystems. UNESCO's MAB Programme and the Project Mesoamerian Biological Corridor by JICA play essential roles in supporting regional conservation efforts. Multilateral land governance and the Central America Integration System (SICA) further enhance cooperation among member countries. A flexible decision-making mechanism based on regional governance ensures effective responses to dynamic environmental challenges. By integrating these initiatives and approaches, Biosphere Reserves can contribute significantly to sustainable development and environmental protection in the Central American region and beyond.

Seventeenth, Dr. Yanuariadi talked about "ITTO's Cooperation on Blue Carbon Ecosystems." The International Tropical Timber Organization (ITTO), established in 1983 under UNCTAD, is the only intergovernmental organization focused on tropical forest resources. It has 75 member countries, including producers and consumers, and aims to promote sustainable management and conservation of tropical forests while encouraging trade in timber from responsibly managed sources. Global challenges, including deforestation and land use changes, threaten the 1.8 billion hectares of tropical forests. Indonesia and Germany have jointly launched the World Mangrove Center (WMC) to address mangrove deforestation and loss, act as a hub for international cooperation, and support healthy mangrove ecosystems and communities amid the climate crisis. Legal entity options include international intergovernmental organization or hybrid models for the center's establishment. The WMC's mission is to provide evidence, solutions, and knowledge exchange for sustainable mangrove management and conservation.

Eighteenth, Mr. Tardieu presented the video on "Sharkfin Bay Project in the Palawan Biosphere Reserve in the Philippines", presenting a 3 min video by Sulubaai Environmental Foundation.

Nineteenth, Dr. Joosten talked about "Blue Carbon and Peatland Ecosystems," discussing 'blue carbon ecosystems' and exploring the link between coastal wetlands and peatland ecosystems. Coastal wetlands, affected by brackish/saline water and tides, are considered peatlands due to their ability to sequester and store peat.

However, the question arises whether they can also be categorized as 'blue carbon' systems. The IPCC emphasizes the need for clear definitions of 'coastal land' and consistent application within countries. Many peatlands worldwide are found in coastal regions and possess significant carbon storage capacity similar to other blue carbon systems (Temmink et al. 2022). These ecosystems are interconnected, with coral reefs aiding mangrove establishment, mangroves stabilizing coastlines, and both filtering nutrients to protect coral reefs. Despite peatlands' high carbon stocks, they are underrepresented in Nationally Determined Contributions (NDCs) compared to mangroves and other blue carbon systems (Elshehawi et al. 2022). The author highlights the importance of understanding the relationship between blue carbon and peatland ecosystems for effective climate change mitigation and adaptation, considering their vital ecosystem services, including water regulation capacity.

Twentieth, Dr. Loughland talked about "Private Sector Financing in research of Blue Carbon, the case of ARAMCO." Aramco, an oil company, is actively involved in extensive mangrove plantations, demonstrating its commitment to environmental sustainability. The project's goal is to restore and expand mangrove ecosystems, which serve as a vital nature-based solution for coastal protection, carbon sequestration, and biodiversity support. Aramco's investment in mangrove restoration includes the creation of an eco-park and a laboratory to study mangrove carbon sequestration and storage. It showcases how private entities can contribute to global environmental challenges and work towards a greener and more resilient future in the Arabian Gulf region.

In the discussion section, we recognized the importance of collaboration and cooperation among experts worldwide to address blue carbon ecosystem issues and mitigate climate change. The conference focused on the Blue Carbon Initiative. The participants recognized the role of higher education in addressing environmental issues and acknowledged the responsibility to engage the younger generation in sustainable practices. We also discussed the potential of blue carbon as a diplomatic tool in international negotiations and emphasized the need for interdisciplinary collaboration across public, private, and academic sectors. In closing, the speakers and organizers highlighted their commitment to continue working together to advance research and action on blue carbon and sustainable development. The conference was seen as a catalyst for further progress in addressing environmental challenges and promoting innovative solutions.

Lastly, Prof. Matsuda, who represented Mr. Kenji Okutsu of Hakkeijima Aquarium, gave a preliminary explanation of the next day's visit about the "Efforts of the Yokohama Blue Carbon Project at Yokohama Hakkeijima Sea Paradise," which aims to promote the use of marine resources, counter global warming, and raise environmental awareness through industry-academia-government collaboration. Through initiatives like seaweed and shellfish consumption, ocean energy utilization, and environmental clean-up, the project aims to create a friendly ocean and reduce carbon emissions. The Yokohama Green Valley Initiative by Yokohama City complements this by focusing on low-carbon energy policies and environmental industries. The Yokohama Blue Carbon Project conducted experiments using

Fig. 1.3 (left) A field trip to Hakkeijima Aquarium "Umi Farm". (Right) Sachiumi's kelp farm in front of Hakkeijima

seawater heat pumps and ship-based power supply, which is probably the first local certificate system of blue carbon offset in the world. Umi Farm, an aquarium for sea education, contributes to awareness-raising among citizens. The project involved transforming the seabed to enhance biodiversity, yielding positive results over time. Collaborative events with Yokohama City engage children in learning about marine conservation. Recognized for its efforts, Hakkeijima Seaparadise received the Minister of the Environment's Commendation. This project demonstrates the effective utilization of marine resources for conservation and sustainability.

On the third day, we visited Hakkeijima Aquarium "Umi Farm". Mr. Kenji Okutsu guided our field trip. We visited the site where we received an explanation the day before. Participants questioned whether the tour was subsidized by the government, but it was run as a commercial enterprise by a privately owned aquarium, and was funded by admission fees from participants. After the Aquarium, we visited a private kelp farm participating in the blue carbon credit program (Fig. 1.3), and after overlooking the only beach in Yokohama City with eelgrass beds and swimming beaches, we visited the Kanazawa Ward Fisheries Cooperative Association in Yokohama City, ending the full-day tour.

1.1 Conclusion

The International Conference on "Blue Carbon Ecosystems for Sustainable Development with Special Emphasis on Mangrove Ecosystems" brought together experts from around the world to discuss the importance of blue carbon ecosystems in sustainable development. These saline ecosystems, like mangroves, play a crucial role in carbon sequestration and soil organic matter accumulation. To achieve the United Nations' sustainability goals, scientific research into blue carbon is vital. The conference highlighted the significance of UNESCO's protected areas, Biosphere Reserves, and Ramsar Wetlands in conserving and engaging with blue carbon ecosystems. Presentations covered global carbon sequestration in mangroves, challenges faced by these ecosystems due to climate change and human activities, and the potential for private sector involvement in conservation efforts. Collaboration among academia, governments, and private sectors was stressed as essential for effective climate change mitigation and protection of blue carbon ecosystems. The conference served as a platform for knowledge exchange and fostering interdisciplinary collaboration towards a greener and more resilient future.

Acknowledgments We thank Drs. Benno Böer, Shahbaz Khan, Antonio Abreu, Luis Aragon, Kiyoshi Fujimoto, Bilquees Gul, Abdul Hameed, Motohiro Hasegawa, Tomomi Inoue, Dai-Yeun Jeong, Hans Joosten, Ryo Kohsaka, Ronald Loughland, Waseem Razzaq Khan, Frederic Tardieu, Martin Zimmer, Kenji Okutsu, Masato Nobutoki for contribution of this conference. We also thank Prof. Emeritus Kaoru Ohno, Dr. Jay Mar D. Quevedo, Dr. Bayan Saideh, and Ms. Aya Horiuchi for editing manuscripts, This chapter is supported by grants by the Research Institute of Environment and Information Sciences of Yokohama National University and UNESCO Chair on Education in Biosphere Reserves for Sustainable Societies.

References

Atwood TB, Connolly RM, Almahasheer H, Carnell PE, Duarte CM, Ewers Lewis CJ, Irigoien X, Kelleway JJ, Lavery PS, Macreadie PI, Serrano O (2017) Global patterns in mangrove soil carbon stocks and losses. Nat Clim Chang 7(7):523–528

Barth HJ, Böer B (eds) (2002) Sabkha Ecosystems: Volume I: The Arabian Peninsula and Adjacent Countries. Springer Science & Business Media

Breithaupt JL, Smoak JM, Smith TJ, Sanders CJ, Hoare A (2012) Organic carbon burial rate in mangrove sediments: strengthening to global budget. Glob Biogeochem Cycles 26:GB3011

Elshehawi S, Kaiser M, Peters J, Joosten H (2022) Opportunities for including peatlands in Nationally Determined Contributions (NDCs) by linking them to mangroves with special attention to the Caribbean region. Proceedings of the Greifswald Mire Centre 03/2022 (self-published, ISSN 2627-910X), 26 p

Hutchison J, Manica A, Swetnam R, Balmford A, Spalding M (2014) Predicting global patterns in mangrove forest biomass. Conserv Lett 7(3):233–240

Inoue T (2019) Carbon sequestration in mangroves: carbon dynamics, policy, and implementation. In: Kuwae T, Hori M (eds) Blue carbon in shallow coastal ecosystems. Springer, Singapore, pp 73–99

Jardine SL, Siikamäki JV (2014 Oct 14) A global predictive model of carbon in mangrove soils. Environ Res Lett 9(10):104013

Lokier SW (2013 Mar) Coastal sabkha preservation in the Arabian gulf. Geoheritage 5(1):11–22

Macreadie PI, Costa MDP, Atwood TB et al (2021) Blue carbon as a natural climate solution. Nat Rev Earth Environ 2:826–839. https://doi.org/10.1038/s43017-021-00224-1

Quevedo JMD, Uchiyama Y, Kohsaka R (2021) A blue carbon ecosystems qualitative assessment applying the DPSIR framework: local perspective of global benefits and contributions. Mar Policy 128:104462

Quevedo JMD, Uchiyama Y, Kohsaka R (2023) Progress of blue carbon research: 12 years of global trends based on content analysis of peer-reviewed and 'gray literature' documents. Ocean Coast Manag 236:106495

Spalding M, Kainume M, Collins L (eds) (2010) World atlas of mangroves, A collaborative project of ITTO, ISME, FAO. Routledge, London

Temmink RJ, Lamers LP, Angelini C, Bouma TJ, Fritz C, van de Koppel J, Lexmond R, Rietkerk M, Silliman BR, Joosten H, van der Heide T (2022) Recovering wetland biogeomorphic feedbacks to restore the world's biotic carbon hotspots. Science 376(6593): eabn1479

Chapter 2
Exploring the Synergy Between Payment for Ecosystem Services and Blue Carbon Credits in Yokohama City

Hiroyuki Matsuda, Masato Nobutoki, and Tomohiro Kuwae

Abstract This paper explores the potential synergies that can arise from the intersection of payments for ecosystem services (PES) and carbon credits. The Yokohama Blue Carbon Project encompasses seagrass beds and seaweed farms, which serve as a prime example of how climate mitigation measures can align with the enhancement of ecosystem functions. Notably, the conservation efforts directed towards seagrass beds preceded the introduction of carbon credits. The local certification of carbon credits, encompassing the valuation of PES, holds the potential to bolster both local nature conservation initiatives and climate mitigation strategies. In cases where PES is dissociated from carbon credits, the beneficiaries of ecosystem services are diverse, and reliance solely on the PES buyer for all maintenance costs is often untenable. Voluntary transactions between sellers and buyers are unlikely to be viable. In contrast, within the realm of carbon credit trading, the mitigation impact remains consistent regardless of the origin of the purchased credits, encouraging buyers to seek the most cost-effective options. Carbon credits enhance the value of seaweed aquaculture operations that could have been established independently of carbon credit incentives. By combining aspects of PES while the Blue Carbon project is based on carbon credits, a local certification system has been not only instituted but also expanded from Yokohama to Fukuoka City and eventually across Japan.

H. Matsuda (✉)
Institute for Multidisciplinary Studies, Yokohama National University, Yokohama, Kanagawa, Japan

M. Nobutoki
Kobe University, Kobe, Hyogo, Japan
e-mail: nobutoki@people.kobe-u.ac.jp

T. Kuwae
Coastal and Estuarine Environment Research Group, Port and Airport Research Institute, Yokosuka, Kanagawa, Japan

Japan Blue Economy Association (JBE), Yokosuka, Japan

M. Clüsener-Godt et al. (eds.), *Blue Carbon Mangrove Ecosystems*, Blue Carbon Ecosystems for Sustainable Development 1,
https://doi.org/10.1007/978-3-031-69553-7_2

Keywords Payments for ecosystem services · Seagrass beds · Seaweed farms · Carbon credits · Voluntary trading

2.1 Introduction

The Yokohama Blue Carbon Project is working to restore and conserve seagrass and seaweed beds, macroalgal beds and macroalgal aquaculture areas, demonstrating the potential to simultaneously achieve climate mitigation and ecosystem function enhancement. In the pursuit of these intertwined goals, two distinct yet interconnected market-based mechanisms have emerged: carbon credits and payments for ecosystem services (PES). Despite their shared commitment to environmental betterment, these mechanisms exhibit fundamental differences in their nature and application.

PES is defined as a voluntary and market-driven approach to environmental conservation and resource management (Milder et al. 2010). However, as discussed below, PES that strictly meet these definitions are not easily established (Vaidhayakarn and Soonthorn Nawaphat 2014). The contribution of beneficiaries to maintaining ecosystem services will be discussed in comparison with mechanisms that are neither voluntary nor market-based, such as environmental taxes.

The overarching aim of this paper is to explore the nuanced interactions between PES and carbon credits and to identify their complementary aspects and potential trade-offs. By examining the synergies between these mechanisms, the joint capacity to promote sustainable development, support local communities and contribute to global climate change goals can be elucidated. This paper delves into the evolution of the Yokohama Blue Carbon Project and reveals how the unique fusion of PES and carbon credits has led to a progressive paradigm in nature conservation and climate change action. Through this analysis, we hope to provide insights into how similar integrated approaches can be used worldwide to address the pressing challenges of our time.

2.2 Relationship Between Payments for Ecosystem Services (PES) and Carbon Credits

PES is a voluntary market-driven initiative that (1) identifies the ecosystem services to be paid for, (2) defines the range of stakeholders—those who receive the benefits and those who conserve or create those benefits, (3) creates a mechanism for how the former will pay the latter, (4) establishes contracts between the two parties, and (5) actually are being maintained as intended, and (6) monitoring and verification work needs to be undertaken (Milder et al. 2010).

Carbon credits and PES can be considered components of corporative social responsibility (CSR), encompassing both charitable and commercial dimensions (Lusch and Vargo 2014). However, carbon offsets are not perceived as charitable actions; rather, they entail compensatory payments for the distinct responsibility of carbon emissions. In this context, carbon offsets and PES might be viewed as instances of creating shared value (CSV) that harmonize the interests of both credit sellers and buyers.

CSV is less prevalent compared to CSR. This disparity may arise due to the numerous challenges in establishing CSV, a predicament similarity faced by PES. (1) it is difficult to evaluate ecosystem services; (2) it is difficult to monitor and verify actual offsetting outcomes; (3) the beneficiaries are diverse, extending beyond payers to encompass all those benefitting from ecosystem conservation. (4) In such a context, it is difficult to establish unambiguous roles for sellers and buyers.

Carbon credits, in contrast, evade certain limitations. Since the outcomes of mitigation measures possess an inherently global character, the entity or nation that absorbs CO2 emissions, along with the method employed, becomes inconsequential as long as an equivalent amount of carbon credits is procured. The necessity for verification extends only to emissions and absorption assessments rather than encompassing the entire spectrum of climate change impacts. As a result, they find their place within the carbon offset market, available for trading.

Carbon credits, in contrast, do not have these limitations. Since the outcomes of mitigation measures possesses an inherently global character, the entity or nation that absorbs CO_2 emissions, along with the method employed, becomes inconsequential as long as an equivalent amount of carbon credits is procured. The necessity for verification extends only to emissions and absorption assessments rather than compassing the entire spectrum of climate change impacts. As a result, they find their place within the carbon offset market, available for trading.

Nevertheless, the precise extent of carbon sequestration fluctuates based on factors within the same forest, necessitating intricate analysis. Conversely, market prices are often established in a broader sense. Furthermore, the calculation parameters are publicly accessible, which might inadvertently result in the generation of 'phantom credits' that adhere to the criteria despite minimal sequestration. Given its fundamentally global scope, the cost of CO_2 offsetting is typically standardized and dictated by market dynamics, irrespective of the expenses tied to carbon sequestration.

As discussed below, Yokohama Blue Carbon extends beyond being a mere carbon credit, encompassing elements of PES. This approach acknowledges not solely the exchange of carbon offset between a seller and a buyer, but also establishes the linkage as a beneficiary and maintainer of ecosystem services. The recognition of participating in the local Yokohama Blue Carbon trading market carries intrinsic value, which introduces dynamic pricing mechanisms.

2.3 Is Climate Change Action Worth the Cost?

Measures to address environmental problems are difficult to implement unless they are actively encouraged in international treaties and other instruments. There are two main reasons for this. (1) Costs are incurred now, but the negative environmental effects will be more pronounced for future generations. The economic impact on future generations takes into account the economic discounting. (2) Climate change strategies encompass both mitigation and adaptation efforts. However, mitigation measures inherently hold a global scope, while adaptation measures yield benefits primarily at the local level. This dynamic can lead to a scenario resembling the 'tragedy of the commons,' where other nations undertake mitigation, leaving a country to potentially gain more from focusing on adaptation initiatives (Nordgren 2016).

Economic discounting and the tragedy of commons are common problems to many environmental issues. It has long been well known in the issue of overfishing in fisheries (Clark 1978). In climate change, four scenarios are available depending on mitigation measures, with scenarios of approximately 4 °C, 3.5 °C, 3 °C and 2 °C rise in radiative forcing (W/m^2) values as RCP8.5, RCP6.0, RCP4.5 and RCP2.6, respectively. For each of these, the future adverse effects of climate change and the future costs of the mitigation measures themselves can be estimated. Assuming a discount rate of 4% per annum in the light of normal market principles, a 4 °C increase would overstate the adverse effects of climate change, but the mitigation costs of the 2 °C scenario would exceed the adverse effects, and the 3.5 °C scenario would minimise the total costs (Lomborg 2020).

However, it is not always appropriate to place the same discount rate on non-market values as on market values. Non-market values are also values that are not traded and for which there are often no substitutes. Adverse effects of climate change include impacts on industrial activities, health impacts and impacts on biodiversity. Increased natural disasters will affect all of these. Of these, health impacts and biodiversity impacts are non-market values and, assuming a discount rate of 0 or 0.1%, the total costs were assessed to be lower for the 3 °C and 2 °C scenarios than for the 3.5 °C scenario (Oda et al. 2023).

Of the five main direct drivers of biodiversity loss—land use change, over-exploitation, climate change, environmental pollution and exotic species, land use change is said to have the greatest impact (Reid et al. 2005), among which the land use change is the biggest contributor to biodiversity deterioration. The Red List of Threatened Species assumes that past rates of decline will continue and that many species are predicted to be extinct or nearly extinct within the next 1 century. Predictions of the impact on biodiversity of climate change issues are often made using species distribution models (SDMs). The current habitat of each species is predicted with explanatory variables such as temperature, precipitation and land use, while future temperature, precipitation and land use are given by the IPCC RCP and SSP (shared socioeconomic pathway) scenarios to predict future habitat area. Impact estimates using SDMs are quantitative and allow scenario analysis, but do not take into account the effects of over-exploitation, pollution and exotic species. However, it is not too bad that it takes into account the biggest factor (land-use change) and climate change.

Mitigation measures involve expanding land for biofuel farms, solar panels and wind farms for increased renewable energy. Mitigation measures involve expanding land for biofuel farms, solar panels, wind power, hydropower, etc. to increase renewable energy.

Mitigation measures involve the expansion of biofuel farmland, solar panels, wind power, hydropower, etc., for increased renewable energy. This will lead to habitat loss associated with land-use change for many organisms. Of these, biofuel sites per unit power generation are more than two orders of magnitude larger than solar power. In the United States, it is estimated that 200,000 km2 will be affected, mainly in temperate deciduous forests and temperate grasslands (McDonald et al. 2009). Large-scale hydroelectric power generation is also concerned about impacts on downstream and upstream ecosystems, in addition to land occupation by dam lakes.

Mitigation measures therefore have positive and negative impacts for biodiversity, in terms of mitigation of climate change impacts and habitat development; the 2 °C scenario has the smallest climate change impacts, but the greatest land-use change impacts of the mitigation measures, with an overall projected reduction in habitat area that is not significantly different from the 3.5 °C target (Hof et al. 2018). It should be noted that this compares the two scenarios as impacts on biodiversity itself, even without considering mitigation costs. Ohashi et al. (2019) showed that the 2 °C scenario was assessed to result in a smaller reduction in habitat area than the 4 °C scenario for all five taxonomic groups mentioned above, even when land-use change was taken into account. However, the 2 °C and 3 °C scenarios were not directly compared.

2.4 The Tragedy of the Mitigation Measures

As mentioned above, mitigation measures were described as the tragedy of the commons, which is often formulated in a prisoner's dilemma game (Matsuda 2021). Furthermore, when developed and developing countries with different economic sizes participate, the non-cooperative solution often leads mathematically to a solution where the richer countries make large public investments (mitigation measures in the context of climate change) and the countries with smaller budget sizes reserve their net income (adaptation measures). This is called the law of equalization in net incomes (Matsuda and Harada 1990). It should be noted that it is not the amount of GDP per capita that is important, but the size of the country's economy as a whole.

However, the non-cooperative solution should not be applied as it is, and the 'tragedy of mitigation measures' should be avoided. The ways to avoid tragedies are (1) to divide common lands into private lands; (2) it would be better if the international treaty secretariat could allocate the amount of effort to realise the optimal solution to each country, which was achieved to some extent in the Kyoto Protocol, which entered into force in 2007, but only agreed by developed countries and the superpowers left the agreement. (3) The Paris Agreement, effective since 2015, mandates that each country must define its own nationally-determined contribution (NDC) and commits to achieving them.

The Paris Agreement, effective since 2015, mandates that each country must define its "nationally-determined contributions" (NDCs) and commits to achieving them. This encourages voluntary contributions while remaining on the commons. (4) As applied to (2) or (3), it sets an upper limit or baseline for the environmental impact (emissions cap) of a member state, a country or a company, and then allows for trading, whereby the excess is purchased from others (Matsuda 2021). Overall targets can be set internationally, as in (2), or voluntarily, as in (3). Carbon credits and PES can also be considered as part of this trading scheme.

2.5 PES Initiatives in Yokohama City and Kanagawa Prefecture

Yokohama City lacks groundwater and other high-quality water sources suitable for drinking and began using the Doshi River as a source of water in 1897. This river is a separate catchment from the Yokohama City. Furthermore, in 1916, Yokohama City purchased the water source forest in Doshi Village and has worked to maintain it. Another notable instance of a city relying on water sources located outside its primary catchment area is evident in New York City (Daily 1997). Approximately 90% of the potable water serving the city's population of ca. nine million is derived from the Catskill Mountains, situated over 150 km away. These regions underwent significant development in the 1980s, resulting in a marked decline in water quality. To address this concern, the city chose to invest between $1 billion and $1.5 billion in the preservation of the Catskill Mountains, as an alternative to embarking on the construction of a filtration plant with a capital outlay of $6 to $8 billion and annual operational costs of approximately $three million. This endeavor stands as a pioneering demonstration of attributing economic value to ecosystem services (Chichilnisky and Heal 1998).

These are similar scenarios with Yokohama City and its water source, Yokohama City has proactively preserved water source forests of Doshi village for over eight decades, predating the efforts of New York City. This form of connection between a city and an external water source, surpassing its catchment boundaries, is labeled an "extended watershed area" (Sadohara 2010). Crucial within this framework is the evaluation of the inherent value within these practices and their translation into scientific knowledge.

Today Yokohama City is the largest basic municipality in Japan with a population of four million and Kanagawa Prefecture, to which Yokohama City belongs, established a Water Source Environment Tax in 2007 (Nakayama et al. 2019). This is referred to "PES -an introduction of good practices in Japan".

However, while the aim is to conserve water sources, it is not a tax based on the amount of water used, but rather a tax that allows the entire population to jointly bear the costs of measures through taxation, taking into account the concept of beneficiary-pays, when not specific residents but a wide range of residents benefit from special environmental conservation measures that go beyond general environmental measures. From this standpoint, equal taxation for all county residents and excess taxation based on income are used in combination. In addition, a sunset clause with a 20-year time limit was introduced on the basis that this taxation system should be verified by 2026. Other prefectures have introduced environmental taxes in recent years, but Kanagawa Prefecture's environmental tax has the highest per capita burden (Kohsaka and Uchiyama 2019). The Japanese Government is also to introduce a forest environment tax from 2024, but this tax is intended to conserve ecosystem services in private forests.

2.6 Yokohama Blue Carbon Project Initiative

Beginning in 2001, efforts have been underway to restore the remaining eelgrass beds within Kanazawa Bay, Yokohama. In 2011, the City of Yokohama acknowledged the carbon sequestration potential of these beds and consequently launched the Blue Carbon (BC) Project. The privately-owned aquarium, Hakkeijima Sea Paradice, has also been involved in the macroalgal bed culture and has started guided tours to show this to the general public. This is carried out as a purely commercial activity. Kelp and wakame macroalgal aquaculture by fishing cooperatives and others is also assessed by Yokohama City for its carbon sequestration effect and is eligible for carbon credit trading (Kuwae et al. 2022). This is a unique example of blue carbon practice in the world for seagrass, macroalgal beds and macroalgal aquaculture.

By trading within the PES framework of voluntary trading between localised sellers and buyers, and because the price is not fixed, the offset price includes not just the carbon price of the carbon credits, but also the value of other ecosystem functions. The Yokohama BC project has been joined by other supporting municipalities, and furthermore, Fukuoka City started another BC project, and in 2020 the Japanese Government approved the foundation of the Japan Blue Economy Association (JBE) as a type of government-approved private cooperative. As long as the status quo is maintained, these projects have become profitable projects without subsidies. Participation in this activity is also expected to contribute to the assessment of ESG investments for companies (Kuwae et al. 2022).

However, there are several challenges. More precise assessment of carbon sequestration capacity is needed, and as it is currently still a local certification, it is not considered a contribution to the Paris Agreement. It also has a PES aspect, but the economic value of ecosystem services is not assessed and relies on the buyer's psychological evaluation. There is a need to identify potential negative impacts of concern if over-planting of algal beds occurs in the future (Kuwae and Hori 2019).

The development of these foundation-led projects has been largely due to the collaboration between industry, government, academia and the private sector, which has seen BC projects as part of community development rather than simply environmental conservation. Also important is the presence of researchers who are energetically involved in both the fundamentals, applications and social contribution, from basic research on carbon storage brains to social implementation of crediting, and even lobbying the IPCC for international standardisation.

2.7 Conclusion

The Yokohama Blue Carbon Project exemplifies the intersection of climate mitigation and ecosystem enhancement through the restoration and conservation of seagrass beds, macroalgal farms, and aquaculture areas. This initiative has embraced the market-driven mechanisms of carbon credits and payments for ecosystem services (PES) to achieve its goals. While both mechanisms share the commitment to environmental betterment, they differ in nature and application.

The intricate relationship between PES and carbon credits would be the potential for synergy and trade-offs. PES, a voluntary market-driven approach, involves diverse stakeholders conserving or creating ecosystem benefits. However, establishing clear roles, evaluating ecosystem services, and verifying outcomes pose challenges. Carbon credits, in contrast, thrive in the globally focused domain of mitigation measures, regardless of their origin. They provide standardized solutions for offsetting carbon emissions, despite limitations in the precision of carbon sequestration measurement.

The impacts of mitigation measures on biodiversity and the costs associated with addressing climate change also give rise to significant concerns. The tragedy of the commons and economic discounting underscore the complexities of climate action, with mitigation and adaptation efforts often yielding varying local and global outcomes. In the economic evaluation of climate strategies, discount rate assumptions for non-market values, such as impacts on health and biodiversity, can greatly influence outcomes.

The Yokohama Blue Carbon Project and similar endeavors highlight the fusion of PES and carbon credits as a progressive paradigm in nature conservation and climate action. Yokohama City's proactive stance towards maintaining water source forests and New York City's investment in the Catskill Mountains demonstrate how cities can secure water from external catchment areas. These integrated approaches offer insights into addressing pressing environmental challenges on a global scale.

Overall, these explorations underscore the complex interplay between economic considerations, environmental conservation, and climate action. It is clear that reconciling economic interests with sustainable practices is essential in shaping a resilient and ecologically balanced future.

Acknowledgement We would like to thank Drs. Taikan Oki, Haruka Ohashi, Takahiro Oda, Kenji Okutsu, Masakazu Hori, Miguel Clüsener-Godt, the UDC members, Yokohama City officials for providing valuable information. This research was funded by the Environment Research and Technology Development Fund of the Environmental Restoration and Conservation Agency of Japan (JPMEERF15S11400) and Yokohama National University.

Conflict of Interest The authors declare no conflicts of interest associated with this manuscript.

References

Chichilnisky G, Heal G (1998) Economic returns from the biosphere. Nature 391(6668):629–630

Clark CW (1978) Mathematical bioeconomics: the optimal management of renewable resources. Wiley

Daily G (ed) (1997) Nature's services: societal dependence on natural ecosystems. Island, Washington DC

Hof C, Voskamp A, Biber MF, Böhning-Gaese K, Engelhardt EK, Niamir A, Willis SG, Hickler T (2018) Bioenergy cropland expansion may offset positive effects of climate change mitigation for global vertebrate diversity. Proc Natl Acad Sci 115(52):13294–13299

Kohsaka R, Uchiyama Y (2019) Forest environmental taxes at multi-layer national and prefectural levels: comparisons of 37 prefectures survey results in Japan. J Jpn For Soc 101:246–252. (In Japanese)

Kuwae T, Hori M (eds) (2019) Blue carbon in shallow coastal eco-systems: carbon dynamics, policy, and implementation. pp 321–46

Kuwae T, Watanabe A, Yoshihara S, Suehiro F, Sugimura Y (2022) Implementation of blue carbon offset crediting for seagrass meadows, macroalgal beds, and macroalgae farming in Japan. Marine policy. 2022 Apr 1;138:104996

Lomborg B (2020) False alarm: how climate change panic costs us trillions, hurts the poor, and fails to fix the planet. Hachette

Lusch RF, Vargo SL (2014) The service-dominant logic of marketing: Dia-log, debate, and directions. Routledge:43–56

Matsuda H (2021) A guideline for ecological risk management Procedure. In: Matsuda H (ed) Ecological risk management: for conservation biol-ogy and ecotoxicology. Springer, pp 271–288

Matsuda H, Harada Y (1990) Evolutionarily stable stalk to spore ratio in cellular slime molds and the law of equalization in net incomes. J Theor Biol 147:329–344

McDonald RI, Fargione J, Kiesecker J, Miller WM, Powell J (2009) Energy sprawl or energy efficiency: climate policy impacts on natural habitat for The United States of America. PLoS One 4(8):e6802

Milder JC, Scherr SJ, Bracer C (2010) Trends and future potential of payment for ecosystem services to alleviate rural poverty in developing countries. Ecol Soc 15(2)

Nakayama K, Shirai M, Yamada M (2019) Effects of environmental taxes on forest conservation: case of the water resources conservation fund in Toyo-ta City. In: Nakayama T, Miyata Y (eds) Theoretical and empirical analysis in environmental economics. Springer, Singapore, pp 49–67

Nordgren A (2016) Climate change and national self-interest. J Agric Environ Ethics 29(6): 1043–1055

Oda T, Takakura J, Tang L, Iizumi T, Itsubo N, Ohashi H, Kiguchi M, Kumano N, Takahashi K, Tanoue M, Tamura M, Zhou Q, Hanasaki N, Hasegawa T, Park C, Hijioka Y, Hirabayashi Y, Fujimori S, Honda Y, Matsui T, Matsuda H, Yokoki H, Oki T (2023/4) Total economic costs of climate change with uneven discount rates for market and non-market values. Environ Res Lett 18 084026

Ohashi H, Hasegawa T, Hirata A, Fujimori S, Takahashi K, Tsuyama I, Nakao K, Kominami Y, Tanaka N, Hijioka Y, Matsui T (2019) Biodiversity can benefit from climate stabilization despite adverse side effects of land-based mitigation. Nat Commun 10(1):5240

Reid WV, Mooney HA, Cropper A, Capistrano D, Carpenter SR, Chopra K, Dasgupta P, Dietz T, Duraiappah AK, Hassan R, Kasperson R (2005) Ecosystems and human well-being-synthesis: a report of the millennium ecosystem assessment. Island Press, p 748

Sadohara S (ed) (2010) Spatio-temporal information platform. Tokyo University Press. ISBN:978-4-13-066852-1, pp 320. (In Japanese)

Vaidhayakarn C, Soonthorn Nawaphat S (2014) Payment for ecosystem services pilot implementation in Mae Sa-Kog ma man and biosphere re-serve. USAID. https://pdf.usaid.gov/pdf_docs/PA00KX8G.pdf

Chapter 3
Blue Carbon Ecosystems for Sustainable Development: Blue Carbon Ecosystems in United Nations Designated Conservation Sites

Bayan Alsaaideh, Yuki Sofue, and Ryo Kohsaka

Abstract Coastal ecosystems, including mangroves, seagrass meadows, and intertidal marshes, in addition to coastal Sabkhas, serve a significant role in carbon sequestration while providing essential services to humans. However, the rapid human activities of coastal ecosystems affect climate, biodiversity, and natural ecosystems worldwide. The purpose of this review chapter is to assess how changes in coastal ecosystems can affect blue carbon storage. For instance, the deforestation of mangroves and the deterioration of coastal Sabkhas can lead to a decrease in blue carbon storage in the coastal areas of Cambodia and Saudi Arabia. Conversely, a study in Algeria has the potential to positively impact blue carbon storage by expanding the Sabkha ecosystem through the development of halophytic vegetation cover. This chapter highlights the importance of remote sensing applications in predicting blue carbon stocks in coastal ecosystems. Firstly, we utilized remote sensing to investigate changes in mangrove extents in Cambodia. Secondly, we reviewed remote sensing studies that investigated the land use/cover changes of coastal Sabkhas in Saudi Arabia and Algeria. These case studies are analyzed to highlight the instrumental value of up-to-date coastal ecosystem extent. Such information enables the monitoring of blue carbon storage at multiple scales, from local to global.

Keywords Blue carbon · Coastal ecosystems · Mangroves · Sabkha · Halophytes · Remote sensing

B. Alsaaideh · Y. Sofue · R. Kohsaka (✉)
Graduate School of Agricultural and Life Sciences, The University of Tokyo, Tokyo, Japan
e-mail: ysofue@g.ecc.u-tokyo.ac.jp

M. Clüsener-Godt et al. (eds.), *Blue Carbon Mangrove Ecosystems*, Blue Carbon Ecosystems for Sustainable Development 1,
https://doi.org/10.1007/978-3-031-69553-7_3

3.1 Introduction

The term "blue carbon" was first used in 2009 in a report created by a collaboration of the United Nations Environment Program (UNEP), the Food and Agriculture Organization of the United Nations (FAO), and the International Oceanographic Commission (IOC)/United Nations Educational, Scientific and Cultural Organization (UNESCO) (Alongi 2018). Different plans have been proposed to improve the protection of blue carbon ecosystems under the United Nations Framework Convention on Climate Change (UNFCCC) to support the Paris Agreement objectives or partially under the Convention on Biological Diversity, Kunming-Montreal Global Biodiversity Framework (Laffoley 2020).

Blue carbon refers to the organic carbon stored by coastal ecosystems, including mangroves, seagrass meadows, and intertidal marshes (Quevedo et al. 2023a). These ecosystems are highly valuable in mitigating and adapting to climate change, as they act as a carbon sink, protect coastlines, and safeguard food security. The scientific community has recently increased its focus on measuring blue carbon due to its natural efficiency (Barnes et al. 2019).

Extensive studies on blue carbon have been conducted in mangrove forests, which have the largest carbon stocks, storing between (212.1 and 725.1) tons per hectare (Beers et al. 2020). However, despite their dynamic and productive coastal ecosystem capable of blue carbon sequestration, less attention has been given to Sabkha ecosystems (Lovelock and Duarte 2019). Coastal Sabkha ecosystems have been identified by various studies as blue carbon ecosystems, which can play a crucial role in monitoring sea level rise, providing data for coastal zone planning, and serving as significant features in arid coastal landscapes (Schile et al. 2017; Al Disi et al. 2023).

In concrete, mapping the extent of carbon values, implementing marine spatial planning and fisheries management measures, and minimizing threats to coastal wetlands can improve the effective management of blue carbon ecosystems. Reflecting these backgrounds, remote sensing techniques for mapping, monitoring, and managing blue carbon habitats can be distinguished as a tool for protecting blue carbon ecosystems at both local and global scales (Malerba et al. 2023).

This chapter reviews the current condition of mangroves and Sabkha, which are vital coastal ecosystems in Cambodia, Saudi Arabia, and Algeria. It estimates changes in mangrove areas and carbon storage over time and analyzes the variable impacts of human activities on the efficiency of these coastal ecosystems for blue carbon storage using remote sensing techniques.

3.2 Allocation of Blue Carbon Ecosystems

Blue carbon ecosystems are distributed on every continent except Antarctica, covering an estimated 49 million hectares (Beers et al. 2020). Mangroves are vital in connecting the carbon cycles of land and sea. They are globally recognized as the

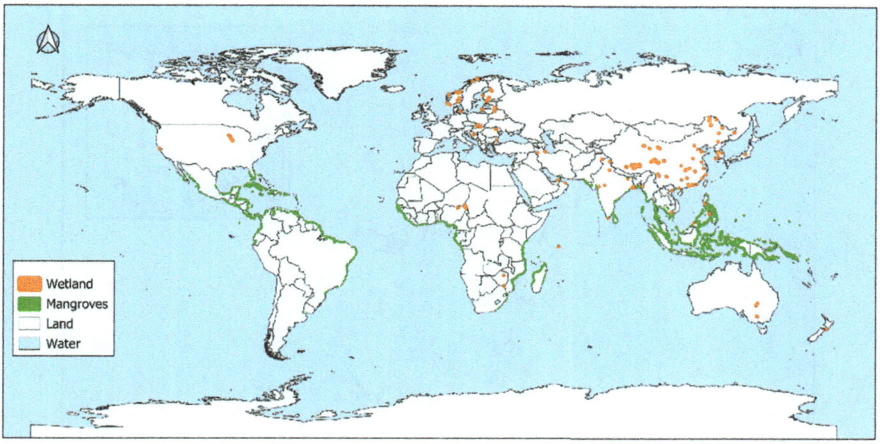

Fig. 3.1 The global distribution of wetlands and mangroves. Source: Global Land Cover by National Mapping Organization (GLCNMO2013) (Kobayashi et al. 2017) and Ramsar website (www.ramsar.org, accessed 2023)

most carbon-rich vegetation due to their high carbon sequestration rate compared to other ecosystems (Zeng et al. 2021). The global distribution of mangroves and wetlands mapped using Global Land Cover by the National Mapping Organization (GLCNMO2013) (Kobayashi et al. 2017) and the Ramsar website (www.ramsar.org, accessed 2023) is shown in Fig. 3.1.

3.3 Mangroves Deforestation Along the Coastline of Cambodia in Southeast Asia, in the Indochinese Peninsula

Cambodia has a high proportion of mangrove forests worldwide, particularly in Southeast Asia. However, the country also experiences some of the highest rates of mangrove deforestation hotspots within the region (Baltezar et al. 2023). Cambodia's mangroves have significant economic and environmental values besides their influence on global climate change. They protect coastlines against high tidal waves and floods, host several animal species, provide food for many marine species, and act as filters of sediments and nutrients in coastal catchments.

3.4 Methodology

This research study focused on identifying the extent of mangrove forests in the southern region of Cambodia. The Landsat TM images from 2006 with a spatial resolution of 30 m and DEM data were used to extract mangroves based on the

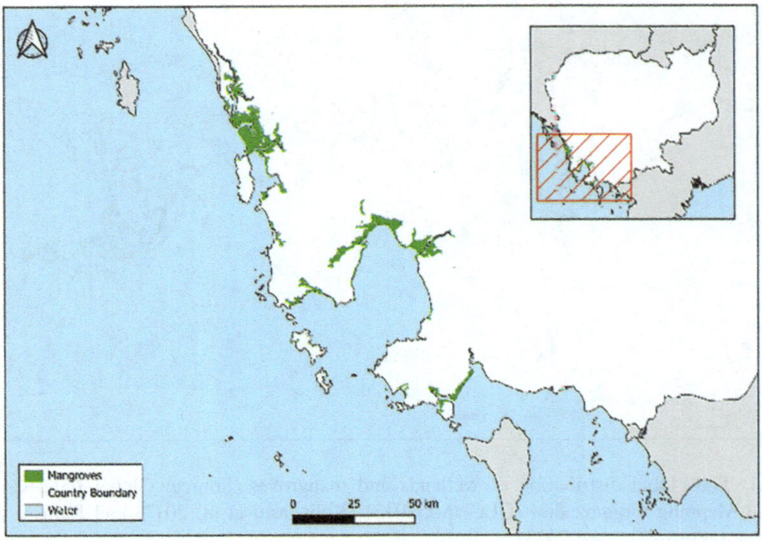

Fig. 3.2 Mangrove forests in Cambodia obtained from classifying Landsat images 2006. Source: Original figure by the author (Bayan) based on Landsat TM

methodology of Alsaaideh et al. (2013). The final classified image displays the distribution of the mangrove forest in Fig. 3.2.

The classified map revealed that fishponds, among other land cover classes (cropland, urban), are the most conspicuous land cover class in the classified map, along with mangrove forests. Figure 3.3 shows how fishponds in the study area were constructed along the mangrove forests.

The accuracy of the resultant map was determined using a confusion matrix with a stratified random sampling method (Tateishi et al. 2011). A stratified random sampling design was used to select points based on the distribution of mangrove forests and other classes shown on the map. According to the research findings, the mangrove forests were distributed within 2 kilometers of the shoreline and below 20 meters above sea level. It was essential to accurately identify the location of these mangrove forests in the study area. By ensuring that all the random samples are distributed within this area, we can obtain a much higher accuracy when mapping out the mangroves. Based on this information, 300 random sample points were selected within the specified threshold values. To verify the accuracy of the resulting map, we utilized Google Earth images to identify the specific locations of the mangroves at the validation sites. Table 3.1 shows the confusion matrix of the mangrove map. The overall accuracy was 91.7%, and the overall kappa value was 0.8; both confirm this classification's high accuracy level and the final map's reliability.

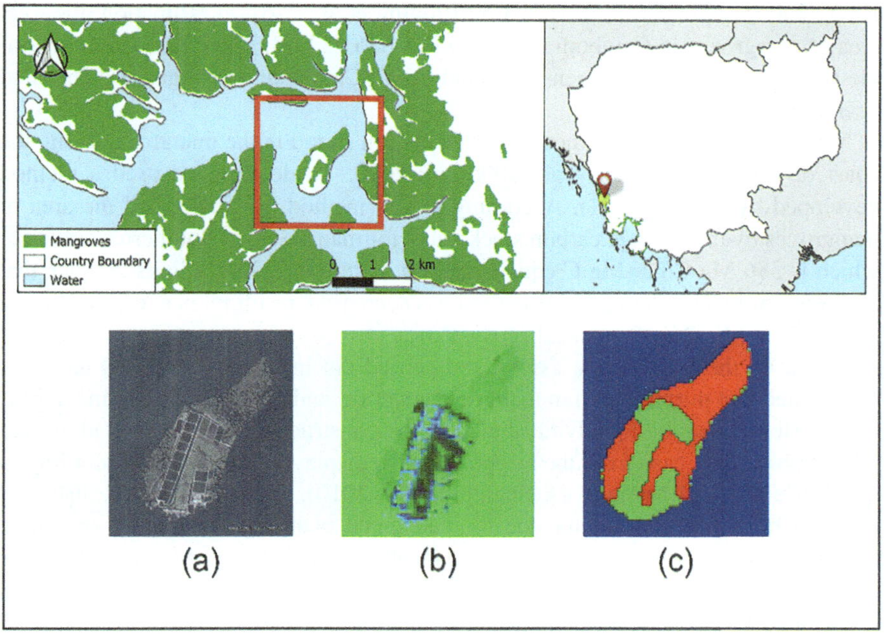

Fig. 3.3 Example of the fishponds that are constructed alongside the mangroves (**a**) The Google Earth image. (**b**) The Landsat TM 2006 and DEM color composite image highlights the mangrove forests in green and the fishponds in blue. (**c**) The classified result indicates the mangrove in red and the fishponds in green. Source: Original figure by the author (Bayan) based on Google Earth image and the classification results of the Landsat TM

Table 3.1 Error matrix of mangrove forests maps derived from Landsat TM in Cambodia

Data	Mangrove	Non-mangrove	Total	User's accuracy (%)
Mangrove	127	23	150	84.7
Non-mangrove	2	148	150	98.7
Total	129	171	300	
Producer's accuracy %	98.4	86.5		

Overall accuracy 91.7%, Kappa statistic 0.8

3.5 Result and Discussion

It is worth noting that the scope of this method used is to estimate the area of mangroves in Cambodia in 2006. It does not include the calculation of degraded mangroves. The analysis revealed that the area covered by mangrove forests was 357.4 km^2. We compared our findings with a study conducted by JICA (2003) on mangroves in Cambodia. According to their report, the estimated existent mangrove area was around 347.08 km^2, and the degraded mangrove area was about 363.039 km^2. Our study's results match those of JICA (2003) concerning the existent mangrove area. Hence, we used the degraded mangrove area estimated by

JICA (2003) in our findings. Based on Bunting et al.'s (2022) estimation, the total area of mangroves in Cambodia by 2020 was 626.92 km^2. Therefore, our comparable results reveal that between 2006 and 2020, Cambodia's mangrove extent had decreased by 93.52 km^2.

We have calculated the amount of blue carbon stored in the mangroves along the coast of Cambodia in the years 2006 and 2020. To do this, we used a method developed by Bastyr (2021). According to this method, we multiplied the area of mangroves by the average carbon stock value for mangroves in Southern Cambodia, which is 386 Mg C/ha (Mg C equals one ton of carbon). Our findings indicate that the extent of mangroves and blue carbon has decreased, resulting in a total decline of 3,609,833.4 Mg C in the coastal ecosystems.

Based on the blue carbon assessment carried out in the study area, it has been determined that the reduction in mangrove coverage and the resulting decline in blue carbon storage can be mainly attributed to the construction of fishponds alongside the mangroves. A case from the research area is displayed in Fig. 3.3. According to the World Atlas of Mangroves (Spalding et al. 2010), the benefits of the fishpond industry have not lasted long due to inadequate planning. The ponds are often abandoned when pollution sets in, which results in the depletion of coastal fisheries and saline pools. This leads to deforestation and reduced ecosystem services, particularly the storage and sequestration of carbon in mangroves.

3.6 Sabkha and Mangrove Areas Located on the Jazan Coast in the Southwest of Saudi Arabia

In order to evaluate the effect of land use/cover (LU/LC) on the storage of blue carbon in coastal ecosystems along the Jazan coast of Saudi Arabia, satellite data from 2009, 2013, and 2021 was utilized (Abd El-Hamid et al. 2023). The LU/LC categories included sabkha and mangroves for all 3 years. As shown in Table 3.2, there was a decline of 27.5 km^2 and 132.87 km^2 in mangroves and Sabkha areas, respectively, from 2009 to 2021. The degradation of large areas of Sabkha and mangroves is mainly caused by increased urban development, which is linked to Saudi Arabia's Vision 2030. Additionally, the study found that soil carbon stocks in the area have significantly decreased since 2013. The ecosystem degradation in the study area could negatively affect carbon storage, resulting in a total loss of −4450675.40 Mg C during the study period (2013–2021). The study suggests that preserving mangroves, sabkha, and vegetation along the coastline can increase

Table 3.2 Area of mangroves and Sabkha during 2009, 2013, and 2021 along the study area of the reference. (Abd El-Hamid et al. 2023)

LU/LC	Area km^2 (2009)	Area km^2 (2013)	Area km^2 (2021)
Mangroves	53.29	28.40	25.79
Sabkha	521.38	460.39	388.51

carbon storage and sequestration. This provides an opportunity for officials to take measures to protect biodiversity, reduce carbon emissions, and preserve soil carbon stocks in coastal wetlands.

3.7 Naama Sabkha of Southwestern Algeria

The nature of Naama Sabkha is known as home to floristic biodiversity and various species of migrating birds, including rare species. The study used remotely sensed vegetation indices and soil sampling analysis to show the growth of Naama Sabkha from (431 ha) in 1985 to (514 ha) in 2018 (Guerine et al. 2020). The results indicate an increase of 83 hectares of the sabkha area throughout the study due to the significant development of dense vegetation cover.

The study by Guerine et al. emphasized the significance of growing halophytes like Salicornia macrostachya and Tamarix gallica in Naama Sabkha. Coastal Sabkha development has the potential to offer valuable resources for agriculture, including food and other economic benefits. This could help promote blue carbon ecosystems and reduce the impact of climate change in arid coastal regions, such as Algeria, which is being used as a case study.

3.8 Conclusion

This review chapter examined the current state of mangroves and Sabkha, two essential coastal ecosystems that store blue carbon, in three different countries: Cambodia, Saudi Arabia, and Algeria. These countries are located in Southeast Asia, West Asia, and Africa, respectively. As the world focuses on climate change mitigation and restoring these ecosystems, it is critical to understand their status.

Comparative analysis conducted in this study determined that human activities such as urban expansion and the development of fish farms have resulted in the decline of Sabkha and mangrove areas in both Saudi Arabia and Cambodia. This loss has had a negative effect on the storage of blue carbon. Additionally, a study (Guerine et al. 2020) has shown that Algeria can increase blue carbon stocks by expanding the Sabkha ecosystem with halophytic plants.

This study suggests utilizing remote sensing techniques to enhance the accuracy of measuring blue carbon. By mapping the ongoing changes in LU/LC of coastal ecosystems, a significant contribution to global environmental conservation can be achieved. This approach provides effective, detailed information on the spatial and temporal aspects of carbon stock. It is crucial to prioritize the protection of blue carbon ecosystems and prevent human interference. The disturbance to these ecosystems can lead to a decline in the amount of carbon sequestered and have a negative impact on human ecosystem services.

Therefore, it is important to consider measures to preserve these ecosystems for the benefit of both the environment and human well-being. In doing so, frameworks such as the DPSIR model, which identifies drivers and pressures for status changes, can be applied in future research (Quevedo et al. 2023b; Kohsaka 2010). Human beings and necessary policy measures can be connected to the drivers and causes of changes through the causal chains in the frameworks. The comparative results presented in this study can be foundations for future scientific explorations and conservation efforts. Studying changes in LU/LC can offer decision-makers a more precise understanding of the status of coastal ecosystems. This information can be used to implement effective management strategies for better monitoring, protection, and restoration of mangroves and sabkha ecosystems worldwide.

Acknowledgements This research is supported and funded by the Asia-Pacific Network for Global Change Research Grant Number CBA2020-05SY-Kohsaka.

References

Abd El-Hamid HT, Eid EM, El-Morsy MHE et al (2023) Benefits of blue carbon stocks in a coastal Jazan ecosystem undergoing land use change. Ecosyst Serv Wetlands 42:1–15

Al Disi ZA, Naja K, Rajendran S et al (2023) Variability of blue carbon storage in arid evaporitic environment of two coastal Sabkhas or mudflats. Sci Rep 13:1–12

Alongi D M (2018) Blue carbon. Springer briefs in climate studies. ISBN 978-3-319-91697-2/ 978-3-319-91698-9 (eBook). Library of Congress control number: 2018944674

Alsaaideh B, Al-Hanbali A, Tateishi R et al (2013) Mangrove forests mapping in the southern part of Japan using Landsat ETM+ with DEM. J Geogr Inf Syst 5:369–377

Baltezar P, Murillo-Sandoval PJ, Cavanaugh KC et al (2023) A regional map of mangrove extent for Myanmar, Thailand, and Cambodia shows losses of 44% by 1996. Front Mar Sci 10:1–16

Barnes DKA, Sands CJ, Richardson A et al (2019) Extremes in benthic ecosystem services, blue carbon natural capital shallower than 1000 m in isolated, small, and young Ascension Island's EEZ. Front Mar Sci 6:1–12

Bastyr A (2021) Mapping 'Blue Carbon' in Cambodian mangroves for the Koh Rong Archipelago. MGEM Project Library. Faculty of Forestry. The University of British Columbia. https://doi.org/10.5683/SP2/DWOXBB, Scholars Portal Data verse, V1

Beers L, Crooks S, Fennessy S (2020) Desktop study of blue carbon ecosystems in Ramsar sites. Report by Silvestrum Climate Associates for the Scientific and Technical Review Panel of the Convention on Wetlands

Bunting P, Rosenqvist A, Hilarides L et al (2022) Global mangrove extent change 1996–2020: global mangrove watch version 3.0. Remote Sens 14:1–32

Guerine L, Belgourari M, Guerinik H (2020) Cartography and diachronic study of the Naama Sabkha (Southwestern Algeria) remotely sensed vegetation index and soil properties. J Rangel Sci 10:172–188

Japan International Cooperation Agency (JICA) (2003) The study on the establishment of GIS Base data. Ministry of public works and transport. The Kingdom of Cambodia. Accessed July 2023

Kobayashi T, Tateishi R, Alsaaideh B et al (2017) Production of global land cover data—GLCNMO2013. J Geogr Geol 9:1–15

Kohsaka R (2010) Developing biodiversity indicators for cities: applying the DPSIR model to Nagoya and integrating social and ecological aspects. Ecol Res 25(5):925–936

Laffoley D (2020) Protecting and effectively managing blue carbon ecosystems to realise the full value to society- a sea of opportunities. An opinion piece by Dan Laffoley for WWF-UK. Woking, Surrey, p 42
Lovelock CE, Duarte CM (2019) Dimensions of blue carbon and emerging perspectives. Biol Lett 15:2–4
Malerba ME, Costa MDP, Friess DA et al (2023) Remote sensing for cost-effective blue carbon accounting. Earth Sci Rev 238:1–16
Quevedo JMD, Uchiyama Y, Kohsaka R (2023a) Progress of blue carbon research: 12 years of global trends based on content analysis of peer-reviewed and 'gray literature' documents. Ocean Coast Manag 236:1–13
Quevedo JMD, Lukman KM, Ulumuddin YI et al (2023b) Applying the DPSIR framework to qualitatively assess the globally important mangrove ecosystems of Indonesia: a review towards evidence-based policymaking approaches. Mar Policy 147. https://doi.org/10.1016/j.marpol.2022.105354
Ramsar Site Database used for wetland mapping. www.ramsar.org. Accessed July 2023
Schile LM, Kauffman JB, Crooks S et al (2017) Limits on carbon sequestration in arid blue carbon ecosystems. Ecol Appl 27:859–874
Spalding M, Kainuma M, Collins L (2010) World Atlas of mangroves. ISBN 978-1-84407-657-4. Routledge, London/Washington, DC
Tateishi R, Uriyangqai B, Al-Bilbisi H et al (2011) Production of global land cover data—GLCNMO. Int J Digital Earth 4(1):22–49
Zeng Y, Friess DA, Sarira TV et al (2021) Global potential and limits of mangrove blue carbon for climate change mitigation. Curr Biol 31:1737–1743

Chapter 4
Perspectives on the Chinese Blue Carbon Credits Market for Safeguarding of Blue Carbon Ecosystems

Shahbaz Khan and Li Li

Abstract This paper focuses on the development of the Chinese blue carbon credit market as a means to protect blue carbon ecosystems. With the rise in carbon emissions and the need to address climate change, carbon credits have emerged as a market-based mechanism to incentivize greenhouse gas reduction. China's commitment to peak emissions by 2030 and achieve net-zero emissions by 2060 has brought attention to carbon markets, and China has significant potential for market-based carbon credits. BCEs (Blue Carbon Ecosystems) offer long-term solutions for mitigating and adaptation of global climate change, managing biodiversity challenges, and creating sustainable development opportunities. The paper discusses the policies and initiatives implemented by the Chinese government to promote the development of the carbon credit market for protecting BCEs. It also explores the challenges and opportunities associated with implementing blue carbon credit trade projects, such as issues related to measurement, reporting, and verification. The Zhanjiang Mangrove National Nature Reserve is used as an example in the paper. This reserve is recognized as a 'Ramsar Wetland of International Importance' and a 'UNESCO Man and Biosphere Reserve'. The example highlights the development of carbon credit mechanisms in China and underscores the importance of integrating various perspectives to ensure the effective functioning of the blue carbon credit market and the long-term protection of blue carbon ecosystems in China.

Keywords Blue carbon ecosystems in China · Conservation and Protection of BCEs in China · Carbon sink and sequestration · Blue economy · Chinese blue carbon credit market

---

S. Khan (✉) · L. Li
UNESCO Regional Office for East Asia, Beijing, China
e-mail: s.khan@unesco.org; l.li@unesco.org

M. Clüsener-Godt et al. (eds.), *Blue Carbon Mangrove Ecosystems*, Blue Carbon Ecosystems for Sustainable Development 1,
https://doi.org/10.1007/978-3-031-69553-7_4

4.1 Background

Blue carbon refers to organic carbon that is captured and sequestered by coastal and marine ecosystems through the carbon cycle (The Global Carbon Budget 2022: https://globalcarbonbudget.org/carbonbudget/; Nellemann et al. 2009). These coastal ecosystems, including mangroves, seagrass meadows and tidal marshes, cover less than 2% of the total ocean area and 0.5% of the seabed, (from the upper intertidal zone down to about 50 m deep, depending on underwater light penetration), but play a significant role in capturing and storing carbon. In fact, they are estimated to store up to 50% of carbon permanently stored in the marine realm. Blue carbon ecosystems (BCEs) are highly productive and manageable. They store carbon within their plants, sediments, and living biomass above the ground, such as leaves, branches, and stems (https://www.thebluecarboninitiative.org/; Herr et al. 2016; Mueller et al. 2019; Jiao et al. 2018). Compared to terrestrial ecosystems, they exhibit faster carbon burial rates and longer carbon storage durations. The waterlogged and anaerobic conditions in these ecosystems prevent carbon decomposition, allowing for extended carbon storage periods. The location of BCEs at the interface of land and sea enables them to accumulate sediment and allochthonous carbon (Alongi 2014; Mcleod et al. 2011; Barbier et al. 2011; Barbier 2016). Recognizing the significance of these coastal carbon sinks, efforts are being made to manage and protect BCEs while addressing gaps in greenhouse gas accounting and management practices.

Recognizing the significance of these coastal carbon sinks, efforts are being made to manage and protect BCEs while addressing gaps in greenhouse gas (GHG) accounting and management practices. BCEs not only serve as cost-effective and scalable natural climate solutions, but they also provide a range of ecological and socio-economic benefits. These include serving as fish habitats, providing shoreline protection against natural disasters, supporting biodiversity, and offering various ecological and socio-economic functions for coastal communities. BCEs also contribute to cultural services such as ecotourism, cultural education and the aesthetic enjoyment of visual and artistic arts. They are an important component of the blue economy and act as a frontline defense against natural hazards, disasters, and human pollution.

4.2 Distribution of BCEs in China

Accurate scoping, monitoring the distribution of BCEs and quantifying the amount of carbon stored in BCEs are great challenges. The global distribution of BCEs is estimated at about 36×10^5–185×10^5 km^2 and potentially storing about 8970–32,650 Tg C^{-1} (Jayathilake and Costello 2018; Bunting et al. 2018; Mcowen et al. 2018), and the carbon burial is 111.4Tg a^{-1} globally (Duarte et al. 2005). Chinese coastal BCEs C storage 0.349–0.835 Tg C^{-1} according to Zhou et al (Zhou et al. 2016). Coastal habitat in China covers an area of 5706 km^2 crossing three climatic zones and possesses mangroves, salt marshes, tidal flats and seagrass meadows (Wang et al. 2021). In China, salt marshes are the main type of BCEs in China and are distributed in the coastal zone of Jiangsu Province, Bohai Bay area and alongside of Chang Jiang River, covering an area of 2979.6 km^2 (Mao et al. 2020). The carbon burial of salt marshes is estimated about 1.19 Tg $C{\cdot}a^{-1}$ (Wang et al. 2021) and the C accumulation rate is 154 g C m^{-2} a^{-1} (Duarte et al. 2005), while the global average is 151 g C m^{-2} a^{-1} (Duarte et al. 2005). In China, mangroves are primarily found in the southern coastal regions. The total mangrove area in China is estimated to be around 258.6 km^2. The distribution area of Mangrove forests in China is small compared with tidal flat and salt marshes. Mangroves are mainly distributed in Guangdong Province Guangxi Province, Hainan Province and Fujian Province of China (Fu et al. 2021), the C accumulation rates of mangrove in China are reported to be 124 g C m^{-2} a^{-1}, while that of the global average is 139 g C m^{-2} a^{-1} (Chen et al. 2021; Chmura et al. 2003). According to Wang et al., the carbon burial of mangroves in China is 0.05 Tg $C{\cdot}a^{-1}$. Seagrass meadows are mainly distributed in Shandong Province in the Yellow Sea, Bohai Sea in Hebei Province and in Hainan Province near the South China Sea covering an area of 94 km^2 (Xu et al. 2021; Wu et al. 2021). The C burial rate is 43 g C m^{-2} a^{-1} which is lower than other BCEs. According to Wang et al., tidal flats are the second largest BCEs in China and the distribution area has been reported from 2374 km^2 to 8588 km^2 by different research groups, Wang et al. found that the carbon burial of tidal flat is more than 0.42 Tg $C{\cdot}a^{-1}$ (Wang et al. 2020a, b) and the C accumulation rate of tidal mudflat is similar with saltmarsh. Liaoning province in the Northeast of China contains the largest area of tidal flats covering 974.73 km^2 (Wang et al. 2021) (Table 4.1).

4.3 The Main Loss Drivers for BCEs in China

Loss and degradation of blue carbon ecosystems (BCEs) have been significant. From the 1980s to 2005, an estimated 20–50% of BCEs worldwide were lost or degraded, and the decline continues at a rate of up to 1–2% per year (Smith 2022). The main drivers of BCEs losses are natural disasters and human unsustainable activities (The United Nations Decade of Ocean Science for Sustainable Development 2021). Data collected by NASA from 2000 to 2016 indicates an increase in the loss of mangrove ecosystems due to natural disasters. South America, Australia, and India have experienced losses driven by natural disasters, while Southeast Asia and Mexico have faced losses primarily due to human activities. Southeast Asia, in particular, has suffered significant loss and damage as mangrove forests have been converted into shrimp ponds, rice areas, and other land uses (Ocean Decade Report, 2021–2022).

China has also experienced a decrease in BCEs due to human activities such as land reclamation, environmental pollution, resource overexploitation and the impacts of climate change. Mangroves have experienced significant losses over

Table 4.1 Distribution and estimated C accumulation rate of coastal wetlands along the coastlines of China

	Distribution Area (km^2)					C accumulation ($Gg\ C \cdot a^{-1}$)			
Province (region) in China	Saltmarshes	Mangrove	Tidal flats	Seagrass Meadow	total	Saltmarshes	Mangrove	Tidal flats	total
Liaoning	974.73		0.01		974.73	162.78			162.78
Hebei	103.47		83.05	45.27	231.78	17.28		13.95	31.23
Tianjin	189.69				189.69	31.68			31.68
Shandong	421.34		342.08	0.50	763.92	70.36		57.47	127.83
Jiangsu	465.98		62.77		528.75	77.82		10.55	88.36
Shanghai	602.66		109.81		712.47	100.64		18.45	119.09
Zhejiang	76.60	1.06	217.40		295.06	12.79	0.21	36.60	49.60
Fujian	51.21	8.27	282.85		342.34	8.55	1.60	48.54	58.70
Guangdong	53.61	92.05	348.07		493.73	8.95	17.86	64.19	91.00
Guangxi	8.98	112.5	697.32		818.81	1.50	21.83	133.94	157.27
Hainan	15.67	36.30	50.31	48.65	150.93	2.62	7.04	9.37	19.02
Taiwan	15.41	7.36	180.75		203.53	2.57	1.43	31.89	35.89
Hongkong	0.02	1.04	0.02		1.09		0.20		0.210
Macau			0.07		0.07			0.01	0.010
Total	2979.4	258.6	2374.5	94.2	5706.7	497.55	50.17	424.96	972.68

Note: The distribution data of saltmarshes, mangroves and tidal flats originated from Mao et al. (Mao et al. 2020). Data on seagrass meadow was from *Xiang* et al. (Xiang et al. 2022), the data of seagrass meadow was collected on the year of 2019. The C accumulation was cited from Wang et al. (Wang et al. 2021) the years. China lost around 40% of its mangrove area and 50% of its tidal marsh area between the 1950s and 2010s (Tong et al. 2023; Xiang et al. 2022).To address the urgent need for conservation and restoration of BCEs, three broad management approaches can be employed: preservation, restoration and afforestation. Implementing these management approaches is crucial for enhancing the sequestration potential of BCEs and mitigating climate change. Conservation and restoration actions should be undertaken at local, national, and international levels to protect and restore these valuable ecosystems (Macreadie et al. 2017).

4.4 China's Commitments to Promote BCEs for Dual Carbon Goals

Under Article 6 of the Paris Agreement, countries are encouraged to include blue carbon into Nationally Determined Contributions (NDCs) to meet their GHG emission reduction goals. Some countries have provided stable policy signals, aligned with funding streams and created relevant incentive strategies to promote the restoration and management of BCEs (Herr and Landis 2016). The IPCC Special Report on the Ocean and Cryosphere in a Changing Climate also confirmed the role of BCEs as nature-based solutions for climate action (https://www.ipcc.ch/srocc/chapter/summary-for-policymakers/). Since June 2021, 33 countries have mentioned 'blue carbon' in their updated NDCs reports (Bin Hu, Natural Resource Economics of China). Countries such as Australia, the United States and China take BCEs as mitigation measurements while small countries in Asia-Pacific regions take BCEs as adaptation measurement ("Blue Carbon and Nationally Determined Contributions." The Blue Carbon Initiative. Accessed June 9, 2022. https://www.thebluecarboninitiative.org/policy-guidance). Currently the United States, Australia and the United Arab Emirates have reported the reduction of GHG emissions due to costal land-use change in their national GHG emission inventories.

Although China has not included blue carbon targets in its national GHG emission inventory yet. Conservation and restoration of BCEs are highlighted in the national issued policies which can support a variety of international commitments. The Law of the People's Republic of China on Marine Environmental Protection, amended in 2017, formally elevated the importance of the protection and restoration of blue carbon resources to the legal level, requiring that the State Council and the local people's governments at all levels along the coasts should take effective measures to protect typical and representative marine ecosystems, such as mangrove forests, coral reefs, coastal wetlands, islands, gulfs, estuaries, and important fishery waters. In 2019, the Communist Party of China (CPC) Central Committee and the State Council issued the Implementation Plan for the National Ecological Civilization Pilot Area (Hainan), which once again pointed out that it is necessary to investigate and study the distribution of blue carbon ecosystems as well as the path and potential for sink enhancement in Hainan Province. The establishment of international carbon emission right trading venues will be explored in compliance with the law. China has launched several blue carbon initiatives with national sustainable development strategies for improving of marine environmental quality and enhancing of marine ecological and environmental governance capacity. The 'Special Action Plan for Mangrove Protection and Restoration (2020-2025)' was issued by the Ministry of Natural Resources and the National Forestry and Grassland Administration of China in 2020, which plans to restore 18,800 hectares of mangrove forests by the end of 2025. Several pilot projects have been implemented in Fujian Province, Guangdong Province and Zhejiang Province by the Chinese government to explore and demonstrate the potential of conserving and restoring coastal ecosystems. In 2022, the National Development and Reform Commission of China and the Ministry of Natural Resources of China jointly issued the 'Key Construction Plan for Protection and Restoration of Costal Ecosystems (2021-2035)', which focuses on the key ecological issues of China's costal belts, it proposes the main objectives of ecological protection and restoration, and coordinates the layout of key projects and key programmes, introduces the construction tasks and the specific initiatives, and clarifies relevant support policies. Blue carbon protection and restoration was highlighted in this state-level strategical plan (Table 4.2).

4.5 The Development of Blue Carbon Credit Market in China

In addition to utilizing national administrative approaches for blue carbon protection, the development of the blue carbon credits market has the potential to make significant contribution to climate change mitigation, biodiversity conservation and sustainable coastal management. It has been mentioned in the UNESCO's report "Custodians of the Globe's Blue Carbon Assets" mentioned that a well-designed blue carbon market mechanism is an effective way to recognize and evaluate the blue carbon assets, providing opportunities to meet the goals of the Paris Agreement and mitigate climate change. Several blue carbon trading projects have been developed globally and verified by the Clean Development Mechanism (CDM,) Verified Carbon Standard (VCS) and Plan Vivo Standard (PVS) (Table 4.3). The reported deal prices of carbon credits from these projects are ranging from US \$5 to US \$35 per t CO_2 e (Zeng 2021) There are several advantages to applying carbon trade markets for blue carbon protection. Firstly, compliance and voluntary carbon credit markets provide price signals for emission reductions that can complement government pricing incentives. This helps to create a market-driven approach to carbon mitigation. Additionally, carbon finance can fill local financing gaps and potentially assist nations facing financial challenges. It can also serve as a revenue source for sustainability initiatives and facilitate blue debt for nature swaps, where countries can receive financial support in exchange for preserving or restoring BCEs.

China launched the world's largest carbon trading market at Shanghai in 2021, however, the blue carbon credit market in China is still in its early stages. According

Table 4.2 The main documentary policies on BCEs in China (Zhao and Wang 2023)

Year	Documents	Authorization entities	Core content
2015	Construction Plan for China's Ecological Civilization Pilot Zone	CPC Central Committee and State Council	Increasing the means of forests, grasslands, wetlands and marine carbon sinks
2015	National Marine Functional Zoning Plan	State Council	Actively develop and utilize marine renewable energy and enhance the function of marine carbon sinks
2016	Work Programme for Controlling Greenhouse Gas Emissions in the 13th Five-Year Plan	State Council	Explore the development of pilot projects for carbon sinks in the ocean and other ecosystems.
2017	Opinions on Improving the Strategy and System of Main Functional Areas	CPC Central Committee and State Council	Exploring the establishment of a blue carbon standard system and trading mechanism.
2017	Concept of Maritime Co-operation for "Belt and Road" Strengthening	Construction National Development and Reform Commission and State Oceanic Administration	International Co-operation on Blue Carbon
2017	The Law of the People's Republic of China on Marine Environmental Protection (Amended)	CPC Central Committee and State Council	Formally elevated the importance of the protection and restoration of blue carbon resources to the legal level
2018	Guiding Opinions of the CPC Central Committee and State Council on Supporting Hainan to Comprehensively Deepen Reform and Openness	CPC Central Committee and State Council	Carry out pilot projects on carbon sinks in marine ecosystems
2019	Implementation Programme of the National Ecological Civilization Pilot Area (Hainan)	General Office of the CPC Central Committee and General Office of the State Council	Carry out pilot projects on carbon sinks in marine ecosystems. Investigate and study the distribution of blue carbon ecosystems in Hainan Province as well as the path and potential for increasing sinks, and carry out different types of carbon sink pilots in some areas.
2020	'Special Action Plan for Mangrove Protection and Restoration (2020–2025)'	Ministry of Natural Resources and the National Forestry and Grassland Administration of Chin	Plans to restore 18,800 hectares of mangrove forests by the end of 2025.

(continued)

Table 4.2 (continued)

Year	Documents	Authorization entities	Core content
2021	Carbon Peak Action Programme by 2030	State Council	Promote the protection and restoration of marine ecosystems as a whole, and enhance the carbon sequestration capacity of mangrove forests, seagrass beds and salt marshes, etc.; and strengthen the research on basic theories, basic methods and cutting-edge subversive technologies for carbon sinks of land and marine ecosystems.
2022	Key Construction Plan for Protection and Restoration of Costal Ecosystems (2021–2035) (2021–2035)	Ministry of Natural Resources of China	This document gives the guideline and suggestion on how to implement regional projects for blue carbon protection and restoration.

to the current law and policy, the China Certified Emission Reductions (CCERs) scheme is an effective supplementary mechanism to the national carbon emissions trading system (ETS), however, blue carbon has not been included in the CCER scheme. The main challenges which hinder the growth of the blue carbon credit market in China (1) lack of standardized methodologies and guidelines of trading mechanisms, the absence of standardized methodologies and guidelines makes it difficult to establish a consistent and transparent framework for trading blue carbon credits. (2) limitations in accurate measurement, monitoring, and verification for carbon sequestration of BCEs, (3) lack of awareness among public, especially among private sectors. Increasing awareness about the importance of blue carbon ecosystems and their potential for carbon sequestration can drive private sector engagement in blue carbon projects.

The Chinese government has recognized the importance of the blue carbon trading market, and has taken steps to establish the blue carbon credit market. In 2021, the Ministry of Natural Resources of China initiated some pilot programmes of baseline study of BCEs, with reference to the methodology recommended by the United Nations Intergovernmental Panel on Climate Change (IPCC) and other international standards. Pilot validation programmes was also carried out in key areas. In 2022, the Ministry of Natural Resources issued the "Marine Carbon Sink Accounting Method", which clarifies the workflow, content, methodology and technology of marine carbon sink accounting and establishes the industry standard.

In 2021, China completed the 'Zhanjiang mangrove afforestation project', which is the first blue carbon trading case in China that meets the Verified Carbon Standard (VCS) and Climate, Community & Biodiversity Standards (CCB). Guangdong Zhanjiang Mangrove National Nature Reserve is located in the coastal mudflats of

Table 4.3 Examples of blue carbon projects verified by CDM, VCS and PVS

Category	Project name	Sponsor	Region	Duration	Project Scope (hm^2)	Estimated Emission Reductions/ (t^{a-1})	Methodology
CDM	Mangrove Rehabilitation Project	Marine Community	Delta of Sinsaloumand Casamance, Senegal	2008.06–2038.06	1700	2704	AR-AMS0003
VCS	Mangrove Rehabilitation Project	Livelihood Carbon Fund	Several deltas and river valleys in the southern and northern parts of Senegal	2009.07–2039.07	No less than 10,000	30,000	AR- AM0014
	Mangrove Rehabilitation Project in Sundarbans, India	Livelihood Carbon Fund	The West Bengal State, India	2010.09–2030.09	No less than 10,000	51,249	AR-AM0014
	Mangrove restoration and coastal green belt protection project	Livelihood Carbon Fund	In the East Coastal Belt of Aceh and North Sumatra, Indonesia	2011.06–2031.06	456,896	124,706	AR- AM0014
	Reforestation and Restoration of Mangrove Forests, Sustainable Livelihoods and Community Development in Myanmar	Worldview International Foundation	Northern Ayeyarwaddy, Myanmar	2015.06–2035.06	2146.48	184,006	AR- AM0014
	Mangrove Forestation in Zhanjiang, China	Ministry of Natural Resources, P. R. China	Zhanjiang, Guangdong Province	2015.04–2055.04	3840 Rehibition	4020	AR- AM0014

(continued)

Table 4.3 (continued)

Category	Project name	Sponsor	Region	Duration	Project Scope (hm^2)	Estimated Emission Reductions/ (t^{a-1})	Methodology
	Blue Carbon Project "Vida Manglar" in the Gulf of Morosquillo, Colombia	Conservation International Foundation	Gulf of Morosquillo, Colombia	2015.05–2045.06	Conservation area 7561 connected water systems	21,310	VM0007
PVS	Mikoko Pamoja, Kenya	Kenya Coastal Zone Ecosystems Association (KCZEA)	Gulf of Gazi, Kenya	2012–2032	Conservation of 107 planting of 10	2482	AR- AM0014
	Madagascar Tahiry Honko Project	Blue Adventure	Bay of Assassin's Bay, South West Madagascar	2017.01–2037.12	Promote the protection and restoration of 1200	1443	AR- AM0014
	Kenya Vanga Project	Kenya Coastal Zone Ecosystems Association (KCZEA)	Southern Coast of Kenya	2019–2039	Stoping mangrove degradation, promote community-based mangrove restoration, and protect 460	5027	VM 0033

Leizhou Peninsula, the southernmost tip of mainland China, with a belt-shaped distribution along the 1500 km coastline of the Leizhou Peninsula, with a total area of 20,278.8 hectares of which the mangrove forests cover an area of 7228 hectares, making it the largest mangrove forest nature reserve in China. In January 2002, the reserve was listed as "Ramsar Wetlands of International Importance", and in 2006, it became one of the first batch of demonstration reserves in China. In 2010, it became a UNESCO Man and Biosphere Reserve of China and a science education base in Guangdong Province in 2022. It is a key area for biodiversity conservation in China and an important base for in situ conservation of international wetland ecosystems. The reserve has 26 species of true mangrove and semi-mangrove plants of 15 families. Zhanjiang has divided the mangrove reserve into a core area, a buffer area and an experimental area to implement different management methods. The core area ensures conservation while the buffer zone allows some teaching and scientific research-related activities. In the experimental area, the Reserve Management Bureau is closely working with the relevant township government, village committee and other units to achieve joint management of aquaculture ponds, and allow moderate aquaculture based on ensuring the normal growth of mangroves (Documentation 2023). The mangrove forest has increased by 127 ha during 2016–2020 which accounts for 59% of national total increases as newly grown mangrove forests were mainly converted from tidal flats and aquaculture ponds (Zhang et al. 2022).The project was developed based on the Clean Development Mechanism (CDM): AR-AM0014 A/R Large-scale Methodology: Afforestation and deforestation of degraded mangrove habitats. The price is about US $9 per t CO_2 e. The project developed the planned 380.4 km^2 of mangrove forest during 2015–2020 in Zhanjiang Mangrove National Nature Reserve, Guangdong Province. The project has generated a total of 5880 t CO_2e emission reductions from the certification, and the emission reductions were used for further mangrove restoration and sustainable development (Chen et al. 2022). A non-government environment protection foundation bought all the carbon credits for offsetting its carbon emissions from organizing business activities. The Zhanjiang mangrove afforestation project provides a model for promoting the realization of the ecological values of mangrove restoration projects in China, and is of positive significance for attracting social funds to invest in mangrove protection and restoration, promoting the development of marine carbon sequestration economy, and achieving carbon neutrality.

With globalization as its main mandate, the Hainan International Carbon Emission Trading Centre was established in 2022 as China's first carbon trade market. It promotes the market-based trading of blue carbon products. In 2023, the first 'blue carbon of salt marshes' transaction was completed in Ningbo, Zhejiang Province, marked the entry of the salt marshes into the market-based trading stage. The transaction was conducted in the form of an auction, with a final price about US $15 per t CO_2e, a premium of more than twice the initial price. The blue carbon credits provide economic opportunities for local communities living near these ecosystems. By engaging in activities such as sustainable aquaculture, ecotourism, or carbon offset projects, communities near Zhanjiang Biosphere Reserve and

Photo 4.1 Mangroves ecosystem in Zhanjiang, Guangdong Province. @zhangjiang.gov.com

Ningbo Biosphere Reserve sites can generate income while ensuring the long-term preservation of blue carbon ecosystems (Photo 4.1).

4.6 Future Perspectives for Functioning BCEs Markets in China

4.6.1 To Provide Robust Policy and Regulatory Frameworks

Consistent policy and regulatory support from the Chinese government will be crucial for functioning BCEs markets and enhancing the willingness of stakeholders. Clear guidelines, regulation and transparent trading mechanisms should be established for fully capitalization of the ecosystem value of BCEs. Efforts should focus on establishing appropriate market mechanisms, standardized technical guidelines, and regulations for blue carbon sink accounting, certification, verification, and trading. This will provide a solid foundation for the development and growth of the blue carbon credit market.

4.6.2 To Improve the Transparent and Accurate GHG Accounting

Strengthen the evidence base (data, maps, etc.) for improved enforcement and mangrove management and enhance the credibility and transparency of BCE markets Accurate measurement and transparent data platform BCEs are essential for the credibility of blue carbon credits (Herr et al. 2016). Monitoring, reporting and verification (MRV) is a basic requirement for addressing any strategy and data obtaining is the biggest challenge for effective management and protection of BCEs. Building on improved data, natural capital accounting supports coastal policies by providing standardized data on the status and economic values of natural assets and how these assets are affected by human activity. The utilization of state-of-art technologies will enhance the process of MRV. Cloud computing and artificial intelligence technologies can be applied to achieve automatic processing, spatial analysis and calculation of blue carbon data, to achieve automatic reporting of greenhouse gas inventory and carbon trading in coastal wetlands at different spatial scales and to meet different precision requirements.

4.6.3 To Raise Public Awareness on BCEs

Increasing public awareness about BCEs is crucial to advocate their roles as nature-based solutions for climate change mitigation and adaptation. Educational campaigns, outreach programs, and community-based initiatives should be promoted to foster a sense of stewardship and responsibility towards these ecosystems. Education for Sustainable Development (ESD) can be customized to strengthen the capacities of potential stakeholders of BCEs. By raising public awareness and understanding, support for the conservation and management of BCEs can be enhanced (Scoullos 2013).

4.6.4 To Enhance the International Collaboration

UNESCO designated sites can serve as valuable demonstration tools for scientific research coordination data standardization as and capacity building to inform sound policy decisions. Strengthen Collaborate with international organizations, research institutions, and other countries to share knowledge, best practices, and experiences in blue carbon conservation and management. Participate in global initiatives and networks focused on blue carbon to leverage international support and expertise. International cooperation and the adoption of certification standards can enhance the credibility and recognition of the blue carbon trading market. Collaboration with other countries and international organizations can facilitate the exchange of experiences and technologies, and enhance the international competitiveness of China's blue carbon trading market.

4.7 Conclusions

The carbon credit market is crucial for valuing the ecosystem services provided by BCEs and can contribute to sustainable financing for blue carbon projects. This financing enables the implementation of large-scale conservation and restoration initiatives, which are essential for the long-term protection of BCEs.

However, there are still challenges that need to be addressed to enable widespread implementation of blue carbon credit projects in China. These challenges include policy barriers, financial constraints, scientific complexities, and social factors. Overcoming these barriers will require concerted efforts from various stakeholders.

Standardized data acquisition and analysis platforms are vital for comparable integrated planning of BCEs. By establishing consistent methodologies and guidelines, it becomes easier to measure and monitor the carbon sequestration and other ecosystem services provided by BCEs. This standardization is essential for the effective functioning of the blue carbon credit market.

UNESCO designated sites, such as the UNESCO Man and Biosphere Reserves and Marine World Heritage sites, can serve as priority sites for testing and applying blue carbon finance schemes. These sites provide valuable opportunities for demonstrating the effectiveness of blue carbon credits in conserving and restoring BCEs. They can also facilitate scientific research coordination, data sharing, policy strategy decision-making, and capacity building for international collaboration.

Overall, the successful implementation of blue carbon credit projects in China requires collaboration and coordination among various stakeholders, including policymakers, financial institutions, scientists, and investors. By addressing the challenges and leveraging the potential of blue carbon credits, it can contribute to the protection and restoration of BCEs while simultaneously addressing climate change and promoting sustainable development.

References

Alongi DM (2014) Carbon cycling and storage in mangrove forests[J]. Ann Rev Mar Sci 6:195–219

Barbier EB (2016) The protective value of estuarine and coastal ecosystem services in a wealth accounting framework. Environ Resour Econ 64(1):37–58. https://doi.org/10.1007/s10640-015-9931-z

Barbier EB et al (2011) The value of estuarine and coastal ecosystem services. Ecol. Monogr 81: 169–193. http://creativecommons.org/licenses/by-sa/3.0/igo/

Bin Hu, Natural Resource Economics of China

"Blue Carbon and Nationally Determined Contributions." The Blue Carbon Initiative. Accessed June 9, 2022. https://www.thebluecarboninitiative.org/policy-guidance.

Bunting P et al (2018) The global mangrove watch — a new 2010 global baseline of mangrove extent. Remote Sens. 10:1669

Chen LZ, Yang CS, Lin GH (2021) The global changes of mangrove. Xiamen University Publishing

Chen GC, Xu J et al (2022) Progress of coastal wetland blue carbon projects in carbon market and advice on facilitating the development of blue carbon projects in China. J. Appl Oceanogr 41(2): 177–184

Chmura GL, Anisfeld SC, Cahoon DR, Lynch JC (2003) Global carbon sequestration in tidal saline wetland soils. Glob Biogeochem Cycles 17(4):1111

Documentation: Open Case Studies/FRST522/2022/The involvement of communities in mangrove management in China: a case study of the Zhanjiang Mangrove National Nature Reserve, Guangdong provincehttps://wiki.ubc.ca/Documentation:Open_Case_Studies/FRST522/2022/The_involvement_of_communities_in_mangrove_management_in_China:_a_case_study_of_the_Zhanjiang_Mangrove_National_Nature_Reserve,_Guangdong_province Accessed on July 30, 2023.

Duarte CM, Middelburg JJ, Caraco N (2005) Major role of marine vegetation on the oceanic carbon cycle. Biogeosciences 2(1):1–8

Fu C, Li Y, Zeng L, Zhang H, Tu C, Zhou Q, Xiong K, Wu J, Duarte CM, Christie P, Luo Y (2021) Stocks and losses of soil organic carbon from Chinese vegetated coastal habitats. Global Change Biol 27(1):202–214

Herr, D., Landis, E. (2016). Coastal blue carbon ecosystems. Opportunities for nationally determined contributions. Policy Brief. Gland/Washington, DC: IUCN/TNC

Herr D, Himes-Cornell A, Laffoley D (2016) National blue carbon policy assessment framework: towards effective management of coastal carbon ecosystems. IUCN, Gland, p 32

https://www.ipcc.ch/srocc/chapter/summary-for-policymakers/

https://www.thebluecarboninitiative.org/

Jayathilake DRM, Costello MJ (2018) A modelled distribution of the seagrass biome. Biol. Conserv. 226:120–126

Jiao N, Wang H, Xu G et al (2018) Blue carbon on the rise: challenges and opportunities[J]. Natl Sci Rev 5:464–468

Macreadie PI et al (2017) Can we manage coastal ecosystems to sequester more blue carbon? Front. Ecol. Environ. 15:206–213. Proposed three key management actions for maximizing blue carbon sequestration within existing coastal vegetated ecosystems

Mao DH, Wang ZM, Du BJ et al (2020) National wetland mapping in China: A new product resulting from object-based and hierarchical classification of Landsat 8 OLI images. ISPRS J Photogramm Remote Sens 164:11–25

Mcleod E, Chmura GL, Bouillon S, Salm R et al (2011) A blueprint for blue carbon: toward an improved understanding of the role of vegetated coastal habitats in sequestering CO2. Front Ecol Environ 9:552–560

Mcowen CJ et al (2018) A global map of saltmarshes. Biodivers. Data J. 5:e11764

Mueller P, Ladiges N, Jack A et al (2019) Assessing the long-term carbon sequestration potential of the semi-natural salt marshes in the European Wadden Sea. ecosphere 10(1):1–15

Nellemann C, Corcoran E, Duarte CM, Valdés L, De Young C, Fonseca L, Grimsditch G (2009) Blue Carbon. A Rapid Response Assessment. United Nations Environment Programme, GRID-Arendal. www.grida.no

Ocean Decade Report, 2021–2022

Scoullos M (2013) Education for sustainable development in biosphere reserves and other designated areas: a resource book for educators in South-Eastern Europe and the Mediterranean. UNESCO Office Venice and Regional Bureau for Science and Culture in Europe (Italy), p 258

Smith JD (2022) Mangrove reforestation provides greater blue carbon benefit than afforestation for mitigating global climate change. Environ Sci J 15(3):123–145. https://doi.org/10.1234/abcd1234

The Global Carbon Budget (2022.): https://globalcarbonbudget.org/carbonbudget/

The United Nations Decade of Ocean Science for Sustainable Development (2021–2030) Implementation Plan. UNESCO (ed.), Paris (IOC Ocean Decade Series, 20.) UNESCO-IOC. 2021.

Tong C, Luo M, Chen L, Zhou J, Huang JF (2023) Methods of carbon sink rate measurement of coastal blue carbon wetland ecosystems, current situation, and challenges in China. Acta Ecologica Sinica 43(17):6937–6950

Wang XX, Xiao XM, Zou ZH et al (2020a) Mapping coastal wetlands of China using time series Landsat images in 2018 and Google Earth Engine. ISPRS J Photogramm Remote Sens 163: 312–326

Wang XX, Xiao XM, Zou ZH et al (2020b) Tracking annual changes of coastal tidal flats in China during 1986–2016 through analyses of Landsat images with Google Earth Engine. Remote Sens Environ 238(110):987

Wang F, Sanders CJ, Santos IR, Tang JW, Schuerch M, Kirwan ML, Kopp RE, Zhu K, Li X, Yuan Z, Liu W, Li Z (2021) Global blue carbon accumulation in tidal wetlands increases with climate change. Nat Sci Rev 8(9):nwaa296

Wu ZJ, Chen SQ, Cai ZF et al (2021) Analysis of distribution change and restoration suggestion of the seagrass beds in Hainan Island[J]. Mar Environ Sci 40(4):542–549

Xiang A, Chuai XW, Li JS (2022) Assessment of the status and capacity of blue carbon in China's coastal provinces[J]. Resour Sci 44(6):1138–1154. https://doi.org/10.18402/resci.2022.06.04

Xu SC, Qiao YL, Xu S et al (2021) Diversity, distribution and conserva-tion of seagrass in coastal waters of the Liaodong Peninsula, North Yellow Sea, northern China: Implications for seagrass conservation [J]. Mar Pollut Bull. https://doi.org/10.1016/j.marpolbul.2021.11226

Zeng, et al. (2021); IFC 2023

Zhang R, Jia M et al (2022) Tracking annual dynamics of mangrove forests in mangrove National Nature Reserves of China based on time series Sentinel-2 imagery during 2016–2020. Int J Appl Earth Obs Geoinf 112:102918

Zhao P, Wang WT (2023) Opportunities, Problems and Countermeasures for China's Blue Carbon Development with Multi-objective Coordination. Eviron. Prot. 51(3):21–24

Zhou CH, Mao QY, Xu X et al (2016) Preliminary analysis of C sequestration potential of blue carbon ecosystems on Chinese coastal zone. Sci Sin Vitae 46:475–486. https://doi.org/10.1360/N052016-00105

Chapter 5
A Desirable Framework for Creating a Carbon-Neutral Society

Dai-Yeun Jeong and Byeongmu Kim

Abstract Climate change is placed at the top among environmental concerns in terms of its impact on nature and humans. Most scholars argue human-induced greenhouse gases, particularly carbon has the most impact on climate change.

As an alternative of the Kyoto Protocol having been promoted from 1997 at a global level, UNFCCC has adopted a strategy named carbon neutrality in 2015 and makes it semi-compulsory for all countries to achieve a carbon neutral society by 2050.

In such a context, this paper aimed at establishing a desirable framework for creating a carbon-neutral society. In order to achieve this objective, this paper examined first the concept and implication of carbon neutrality, and then established the direction and content for creating a carbon-neutral society including a framework to approach carbon neutrality, approaches towards a policy for achieving carbon neutrality, introduction of governance to policy-making process, and policy effect analysis for deciding what policy has a priority to be launched.

The Concept and Implications of Carbon neutrality: Carbon neutrality is defined as removing as much carbon from the atmosphere as we put in. This results in net-zero carbon emissions. The overall goal of carbon neutrality is to achieve a zero-carbon footprint. If carbon neutrality is successful, current climate change will continue, but does not get more serious because no more additional human-induced carbon will be emitted into the atmosphere.

A Framework to Approach Carbon Neutrality: The framework should be established in two ways. One is to societal system, and the other one is to climate

D.-Y. Jeong (✉)
Environmental Sociology, Director of Asia Climate Change Education Center, Jeju, South Korea

Jeju National University, Jeju, South Korea
e-mail: jeongdy@jejunu.ac.kr

B. Kim
Environmental Sociology, Jeju Carbon Neutrality Research and Support, Jeju, South Korea
e-mail: kbm@jri.re.kr

M. Clüsener-Godt et al. (eds.), *Blue Carbon Mangrove Ecosystems*, Blue Carbon Ecosystems for Sustainable Development 1,
https://doi.org/10.1007/978-3-031-69553-7_5

change impacts. The former is for reducing the driving-force determining climate change impact. The latter is for improving the current status of climate change.

Approaches towards a Policy for Achieving Carbon Neutrality: Three approaches should be included. They are the nature-based, the technology-based, and the societal system-based approach.

Introduction of Governance to Policy-Making Process: This is a necessary means to ease off the social conflict through internationalization of social conflict in advance. In other words, the introduction of governance is for establishing policies based on social consensus.

Policy Effect Analysis for Deciding What Policy Has a Priority to Be Launched: A great number of carbon neutrality related policies can be established in the policy-making process. All of them can't be implemented because a lot of finances are required to implement policies. Therefore, it is inevitable to implement a limited number of policies within the availability of finance. This is the decision of policy priority that are to be implemented, and the policies with relatively higher priority should be selected and implemented.

Keywords Concept of low-carbon · Concept of carbon-neutrality · Concept of carbon-free (carbon-zero) · Existing approaches to carbon reduction · Framework of carbon-neutral society · Capacity building for achieving carbon-neutral society

5.1 Introduction

Climate change is an environmental problem, that is placed at the top among environmental concerns. The reason for this is that the impact of climate change on nature and humans is most serious comparing to other environmental problems. There are two groups of scholars regarding the causes of climate change (Jeong 2004: 241–245). One of the scholars is arguing that the change is based on natural factors, and the one is arguing that the change is human induced via greenhouse gases having been emitted since the eighteenth century when industrialization started to be promoted. The latter is the major argument, and as we know, the Kyoto Protocol stipulated six greenhouse gases in 1997.

The six human-induced greenhouse gases have different impact on climate change in terms of per ton or as an emission unit and total quantity emitted. For example, the impact of methane is about 21 times higher than carbon per ton, while the impact of carbon is higher than methane in terms of total quantity emitted at a global level.

Most scholars argue that carbon has the most impact on climate change in terms of the quantity emitted among the six human-induced greenhouse gases. Two keyways in which these gases differ from each other are their ability to absorb energy (their radiative efficiency), and how long they stay in the atmosphere (their lifetime). Summarizing the different arguments by scholars, carbon accounts for about 76% of total greenhouse gas emissions and occupies the effect on climate change by 68% at minimum and 98% at maximum. This is the reason why carbon is emphasized when climate change is mentioned, even though there are six human-greenhouse gases stipulated by the Kyoto Protocol.

In such a context, the Kyoto Protocol which is based on a low-carbon strategy responding to climate change at a global level has served as a basic response strategy to climate change for all countries. The Kyoto Protocol contributed to reducing the increased rate of carbon emissions but resulted in low contribution to mitigating climate change. As an alternative of the Kyoto Protocol, UNFCCC (United Nations Framework Convention on Climate Change) has adopted a new strategy named the Paris Agreement at COP21 in 2015. The Paris Agreement promotes carbon neutrality and makes it semi-compulsory for all countries to achieve a carbon neutral society by 2050.

In such a context, this paper aims at establishing a desirable framework for creating a carbon-neutral society. In order to achieve this objective, this paper will examine first the concept and implication of carbon neutrality and will be followed by a direction and content for achieving carbon neutrality.

5.2 The Concept and Implications of Carbon Neutrality

There are three similar and interrelated terminologies, but different concept. They are low carbon, carbon neutrality, carbon-zero or carbon-free. Their concepts and implications are as below.

5.2.1 Low Carbon

The concept of low carbon is defined differently by scholar (eg. Skea and Nishioka 2008; Fu et al. 2010; Mulugetta and Urban 2010; Yuan et al. 2011; Rosenbloom 2017), but is summarized as below.

Low carbon simply means less carbon. Low carbon has its root in the UNFCCC adopted in the Rio Environment Summit held in 1992 and is generally used to describe forward-looking national economic development plans or strategies that encompass low emissions and/or climate-resilient economic growth. Low carbon includes no absolute level of reduction, but is promoted just such as reduction by 10% or the way to less emissions, etc. Low carbon also includes provisions to reduce vulnerability to the impact of climate change.

Thus, if low carbon is promoted, current climate change will continue, but its acceleration will be reduced.

5.2.2 Carbon Neutrality

The terminology of carbon neutrality has been around for a long time with a different angle on its concept and implications (e.g. CNCA 2014; Can 2021; Huovila et al. 2022; Xianhua et al. 2022). Carbon neutrality is defined as removing as much carbon from the atmosphere as we put in. This results in net-zero carbon emissions. The overall goal of carbon neutrality is to achieve a zero-carbon footprint.

The approach to zero-carbon footprint includes carbon offsets (e.g. wind farms, solar parks), carbon sinks, buying enough carbon credits to make up the difference if the carbon is removed at the same rate as its being emitted, and industrial processes such as production of carbon-neutral fuel, etc. In other words, true carbon neutrality only exists at the planetary scale when there's a balance between the greenhouse gasses the world produces and the removal of those gasses from the atmosphere, primarily through the approaches mentioned above.

If carbon neutrality is successful, current climate change will continue, but does not get more serious because no more additional human-induced carbon will be emitted into the atmosphere.

5.2.3 *Carbon-Zero*

Carbon-zero (also sometimes described as carbon-free, zero-emission or zero-carbon) is used as the same meaning but is complicated by the fact that there's no universally accepted definition for the term, and it loses some of its substance when it's applied to the empirical reality. According to the meaning of terminology, carbon-zero means exactly what it says: the business or activity produces no greenhouse gas emissions. However, actual carbon-zero is not possible so long as humans are active in life. Therefore, the concept and implications of carbon-zero are summarized synthetically as below (e.g. McLenan et al. 2014; Choi et al. 2019; HSBC 2019; Can 2021; Rong et al. 2021).

Carbon-zero means removing more carbon than humans emit. Considering that 'actual carbon-zero' is not possible, the best that we can achieve is 'virtual zero emission' (at least a 90% reduction) and/or 'negative carbon emissions' (artificial carbon sink by tree planting, carbon capture and storage, etc.). Therefore 'true carbon-zero' is defined as removing carbon more than we emit through (virtual zero carbon) + (some negative carbon).

If carbon-zero is successful, current climate change will be gradually reduced, and moves toward climate neutrality which is defined as the status of the climate before industrialization in the eighteenth century.

5.3 A Direction and Content for Achieving Carbon Neutrality

What factors should be included for achieving carbon neutrality, and how should the direction and content of the factors be established? An efficient and effective answer to this question will become a desirable framework for achieving carbon neutrality.

At the least, four factors should be included in establishing a carbon neutral society. They are a framework to approach carbon neutrality, approaches to policy for achieving carbon neutrality, introduction of governance to policy-making process, and policy effect analysis for deciding what policy has a priority to be launched.

5.3.1 A Framework to Approach Carbon Neutrality

The first step is how to approach carbon neutrality. As is proved empirically, societal systems being profiled as industrialization, urbanization, consumerism, and globalization is the cause of emitting human-induced greenhouse gases, and climate change impacts on nature and humans/society. In other words, societal systems are the driving-force determining the status of climate change and its impact on nature and humans/society.

Therefore, the approach to carbon neutrality should be applied in two ways. One is to societal system, and the other one is to climate change impacts. The former is for reducing the driving-force determining climate change impact. The latter is for improving the current status of climate change. In other words, the former is the strategy for reducing the emission of greenhouse gases, while the latter is the strategy for treating the greenhouse gases emitted from societal systems. Such a framework is diagramed Fig. 5.1 (Jeong 2020).

5.3.2 Approaches Towards a Policy for Achieving Carbon Neutrality

A great number of specific policies can be derived from the two directions of strategies on carbon neutrality in Fig. 5.1. However, the specific policies can be

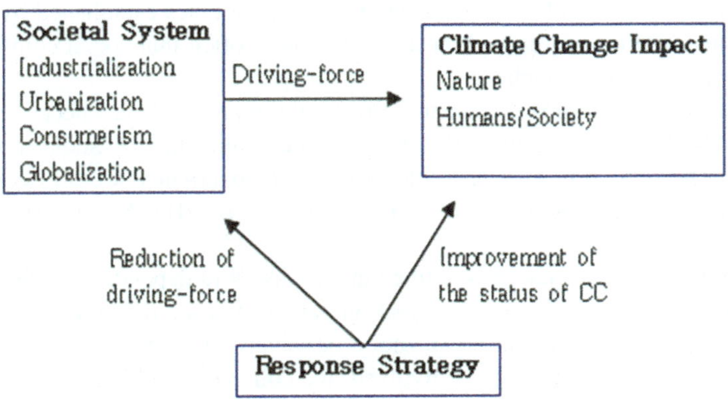

Fig. 5.1 A framework to approach carbon neutrality

broadly categorized into three categories. They are nature-based approaches (Eggermont et al. 2015; IUCN 2016; Kabish et al. 2017; Jeong 2020), technology-based approaches (Mankins 2009; Jeong 2020), and societal system-based approaches (Jeong 2020).

The nature-based approach is the solution through the action to protect, sustainably manage, and restore natural and modified ecosystems that address societal challenges effectively and adaptively, simultaneously benefiting people and nature. The examples include creating the sources of carbon sinks, expanding urban space, and expanding blue carbon, etc.

The technology-based approach is the solution via technology helping to improve the efficiency and effectiveness of achieving carbon neutrality. The examples include improvement of energy efficiency, supply of new and renewable energy, carbon capture/storage/sequestration, and carbon utilization, etc.

The societal system-based approach is a solution through transforming the existing societal system to a low carbon socio-economic system. The transformation of economic systems would be though maximizing profit as the goal, minimizing ecological costs should therefore be considered with the same priority. Citizens' lifestyle should be transformed to eco-friendly behavior, and cultural ethos should be transformed from consumerism to environmentalism.

In detail, there are many sectors to be transformed to new systems. The sectors are land-use, transportation, green space, building, ecosystem, energy, living environment, and waste, etc. Considering such broad sectors as the target of transformation, natured-based and technology-based approaches are a means necessary for societal system-based approaches.

5.3.3 *Introduction of Governance to the Policy-Making Process*

After establishing the approaches for achieving carbon neutrality, the concrete policies of carbon neutrality to be launched should be established. Government is the major agent to establish policies. However, the policies established by government alone are not good enough because it is quite often that social conflict arises when the policies are launched.

The introduction of governance to policy-making processes is a necessary means to ease off the social conflict through internationalization of social conflict in advance. The definition and desirable process of governance would be as below for establishing the policies of carbon neutrality (Hufty 2011; Speer 2012; Rotberg 2014; Grindle 2017).

The concept of governance is defined in various ways depending on the target to which governance is applied. In general, governance is the process of making and enforcing decisions within an organization or society. It is the process of choosing the right course among the actors involved in a collective problem that leads to the creation, reinforcement, or reproduction of acceptable conduct and social order. In addition, good governance means that processes and institutions produce results that meet the needs of society while making the best use of resources at their disposal.

As explained above, the purpose of governance is for reducing social conflict in the process of implementing policies through internalization of social conflict in advance. In other words, the introduction of governance is for establishing policies based on social consensus. The desirable participants in governance are expert, civil organization, resident, and stakeholder, etc.

5.3.4 *Policy Effect Analysis for Deciding Policy Priorities to Be Launched*

A great number of carbon neutrality related policies can be established in the policy-making process. All of them can't be implemented because a lot of finances are required to implement policies. Therefore, it is inevitable to implement a limited number of policies within the availability of finance. This is the decision of policy priority that are to be implemented, and the policies with relatively higher priority should be selected and implemented. Such analysis of policy priority is a category of policy effect analysis being conducted for selecting the final policies to be launched (Walker et al. 2001; Jeong 2012; Friedman 2017).

Table 5.1 is an example of policy effect analysis in terms of investment efficiency (Jeong 2012). As is identified from <Table 5.1>, Substitute of Traffic Signal with LED is relatively most effective in terms of budget for reducing 1 ton of carbon emission and followed by Supply of Clean Energy and Afforestation.

5.4 Concluding Remarks

The objective of this paper was to establish a desirable framework for creating a carbon-neutral society. In relation to the objective, this paper examined the concept and implications of carbon neutrality comparing low carbon and carbon-zero and followed by the direction and contents to be included in the framework.

Table 5.1 An example of policy effect analysis in terms of investment efficiency—for the case of Jeju Province, South Korea

Policy to be implemented	Budget (million in US$)	Reduction quantity (ton)	Budget for reducing 1 ton (US$)
Afforestation (6500 ha)	93	38,350	2430
Supply of Clean Energy	104	227,372	460
Substitute of Traffic Signal with LED	3	9735	310

The framework focuses on policymaking which is the role of government in realizing carbon neutrality. Enterprise, civil organization, and citizen are also the agents responding to climate change, so the framework of their roles in creating a carbon-neutral societies should also be established.

The environmental problems including climate change are a side-effect from industrialization. In this sense, we need to examine the implication of industrialization. During the agricultural society 300 years ago, humans were economically very poor and everyday life was very inconvenient, while nature enjoyed its original quality. In order to overcome the economic poverty and inconvenience in everyday life, industrialization has been advanced since the eighteenth century.

It is no doubt that industrialization has resulted in improvement of material affluence and convenience in life. These results are being achieved from an increase in extraction of natural resources and an increase in emission of gaseous/liquid/solid waste, all of which are the causes of nature being polluted and/or destroyed. The pollution and/or destruction of nature is the crisis of nature which is linked to the crisis of human existence on the earth. This is because nature can exist without humans, but human can't survive without nature.

In this sense, it may be argued that humans are the beneficiary and victim of industrialization. This means that industrialization is a self-contradiction that humans started to commit from the eighteenth century. In essence, all human activities respond to environmental problems including climate change, and are the actions for solving this self-contradiction. We are now required to examine if humans can continue to improve their material affluence and convenient way of life.

References

Can W (2021) The road to low-carbon transition under the vision of carbon neutrality. China Environ Manage 13(1):13–15

Choi SA et al (2019) Flying with climate liability? Economic valuation of voluntary carbon offsets using forced choices. Transp Res Part D: Transp Environ 62:225–235

CNCA (Carbon Neutral Cities Alliance) (2014) Framework for Long-term Deep Carbon Reduction Planning

Eggermont H et al (2015) Nature-based solutions: new influence for environmental management and research in Europe. Ecol Perspect Sci Soc 24(4):243–248

Friedman LS (ed) (2017) Does policy analysis matter?: exploring its effectiveness in theory and practice. University of California Press

Fu JF et al (2010) Conceptual identification and evaluation index system for low carbon economy. China Popul Resour Environ 20(8):38–43

Grindle MS (2017) Good governance: a critique and an alternative. Governance 30(1):17–22

HSBC Center of sustainable finance (2019) Towards the Zero-Carbon City

Hufty M (2011) Investigating policy processes: the governance analytical framework. In: Wiesmann U et al (eds) Research for sustainable development: foundations, experiences, and perspectives. Geographical Bernensia, Bern, pp 403–424

Huovila A et al (2022) Carbon-neutral cities: critical review of theory and practice. J Clean Prod 341:130912

IUCN (International Union for Conservation of Nature) (2016) In: Cohen-Shacham E, Walters G, Janzen C, Maginnis S (eds) Nature-based solutions to address global societal challenges

Jeong DY (2004) Environmental sociology. Acarnet, Seoul

Jeong DY (2012) An effectiveness analysis of climate change policy in South Korea. J Environ Impact Assess (Korean Society of Environmental Impact Assessment) 20(5):585–600

Jeong DY (2020) An approach to carbon-free society as a response to climate change. Paper presented at International Conference on Carbon Dioxide Utilization Summit organized by Active Communications International and held in Orando, USA. on February 26–27, 2020

Kabish N et al (eds) (2017) Nature-based solutions to climate change adaptation in urban areas. Springer

Mankins JC (2009) Technology readiness assessments: a retrospective. Acta Astronaut 65(9–10): 1216–1223

McLenan JC et al (2014) Voluntary carbon offsetting: who does it? Tour Manager 45:194–198

Mulugetta Y, Urban F (2010) Deliberating on low carbon development. Energy Policy 38: 7546–7549

Rong J et al (2021) Influence of carbon market on carbon capture, utilization and storage industrialization development. Therm Power Gener 50:43–46

Rosenbloom D (2017) Pathways: an emerging concept for the theory and governance of low-carbon transitions. Glob Environ Chang 43:37–50

Rotberg R (2014) Good governance means performance and results. Governance 27(3):511–518

Skea J, Nishioka S (2008) Policies and practices for a low-carbon society. Clim Pol 12:S5–S16

Speer J (2012) Participatory governance reform: a good strategy for increasing government responsiveness and improving public services? World Dev 40(12):2379–2398

Walker WE et al (2001) Adaptive policies, policy analysis, and policy-making. Eur J Oper Res 128(2):282–289

Xianhua W et al (2022) A Review of the theoretical research and practical progress of carbon neutrality. Sustain Oper Comput 3:54–66

Yuan H et al (2011) What is low-carbon development? a conceptual analysis. Energy Prodedia 5:1706–1712

Chapter 6
Sabkha Ecosystems as Blue Carbon Reserves

Abdul Hameed, Benno Böer, Miguel Clüsener-Godt, and Bilquees Gul

Abstract The term ***Sabkha*** refers to coastal and inland salt flats found in arid or semi-arid regions, where the evaporation of saline water leaves the surfaces covered in salt. Significant levels of organic carbon are found in the sediments of Sabkhat. This carbon is reportedly sequestered in a variety of ways, including through the photosynthetic carbon gain by Sabkha halophyte vegetation, including micro-algae, accumulation of organic matter through nearby effluents, formation of carbonate minerals, and activity of mat-forming blue-green algae and microorganisms, among others. Therefore, sabkhat, particularly those in coastal settings, may be thought of as a significant blue carbon accumulator. Along with numerous additional economic advantages (such as salt production and geo-tourism), sabkhat also provide ecological services including nitrogen cycling and shoreline protection. However, sabkhat are threatened by human activities, such as urbanization, industrialization, land reclamation, groundwater extraction, and pollution, thereby warranting serious protection and conservation efforts. This article summarizes the research, biodiversity, ecological importance and conservation status of sabkhat with respect to carbon sequestration and climate change.

Keywords Arid region · Coastal area · Halophyte · Sabkha · Salinity · Salt flat

---

A. Hameed (✉) · B. Gul
Dr. Muhammad Ajmal Khan Institute of Sustainable Halophyte Utilization,
University of Karachi, Karachi, Pakistan
e-mail: ahameed@uok.edu.pk; bilqueesgul@uok.edu.pk

B. Böer
UNESCO New Delhi Office in South Asia for Bangladesh, Bhutan, India, Nepal, the Maldives, and Sri Lanka, New Delhi, India
e-mail: b.boer@unesco.org

M. Clüsener-Godt
UNESCO Chair on Biodiversity Safeguard for Sustainable Development, CFE,
Centre for Functional Ecology – Science for People & the Planet, Department of Life Sciences,
University of Coimbra, Coimbra, Portugal
e-mail: clusenergodt@uc.pt

M. Clüsener-Godt et al. (eds.), *Blue Carbon Mangrove Ecosystems*, Blue Carbon Ecosystems for Sustainable Development 1,
https://doi.org/10.1007/978-3-031-69553-7_6

6.1 Introduction—What Is Sabkha?

Sabkha or sebkha (سبخة) is an Arabic term generally used for the salt-crusted coastal flats. It also includes 'inland flats' mostly in arid or semi-arid climates, where evaporite-saline minerals accumulate at the surface (Fig. 6.1; Thomas 1929; Harwood 1993; Goodall and Al-Belushi 2020; Barth and Böer 2002a). Sabkhat (سبخات) is the correct plural of sabkha, despite the frequent use of sabkhas (Neal 1975). There are many similar terminologies in other languages used for salt-flats, which include *pan* (South Africa), *salt pan* (Austraila), *nor* (Mongolia), *salar* (Chile), *kavir* (Iran), *takyr* (Russia), *chott, shott,* or *shatt* (Maghreb) and *playa* (USA) (Barth and Böer 2002a). However, these terminologies may not always coincide with the geological definition(s) of sabkha and care should be taken when using these terminologies.

In the scientific literature, the term sabkha has been variously defined. Some authors such as Philby (1933) and Thesiger (1946) restricted it to the coastal flats. Whereas others like Kinsman (1969), Glennie (1970) and Whitten and Brooks (1972) differentiated it into coastal and inland sabkha. Briere (2000) defined sabkha as "a shallow basin limited to marginal marine settings and associated with several per cent gypsum or gypsum parting laminae due to preferential halite dissolution during flooding". Thereby, Briere (2000) restricted this terminology to the coastal areas and preferred the term 'playa' for the inland/continental Sabkha. However, Barth (2001) disagreed and differentiated playas and inland sabkhat according to their geological origin. According to Barth (2001) inland sabkhat are "located in large desert plains within unconsolidated sediments. They are not catchments of internal drainage areas like playas and flooding occurs only episodically and exclusively by precipitation". Hence, inland sabkhat were distinguished from the playas in their aeolian nature.

In 2002, Barth and Böer in the first volume of 'Sabkha Ecosystems' defined sabkhat based on geomorphic features as 'salt encrusted periodically flooded coastal as well as inland flats' and classified sabkhat into the following landform categories: (1) Coastal Sabkha, which are frequently inundated by marine tides; (2) Supralittoral

Fig. 6.1 Examples of Sabkha ecosystems. Panel A is a coastal sabkha located along Hawksbay beach Karachi, Pakistan, and Panel B is an inland sabkha located in the tharparker district of Sindh, Pakistan, both with vegetated hummocks

Sabkha, which are less-frequently inundated by marine tides (i.e. intermediate between 1 and 3); and (3) Inland Sabkha, which are never inundated by marine tides but are flooded episodically by precipitation. Furthermore, Barth and Böer (2002a, b) also classified sabkhat based on inundation regimes (i.e. permanently wet, periodically flooded and permanently dry) and anthropogenic activities (i.e. low saline, landfill, excavation, plantation and other), thereby making their sabkhat description comprehensive and applicable to various purposes. The terminology of Sabkha is now gaining popularity, which can easily be gauged by the increasing number of research articles. The aim of this review is to provide an overview on research, diversity, and threats to Sabkha ecosystems. Furthermore, the importance of sabkhat as a potential blue carbon reserve of the future will also be discussed.

6.2 Characteristics of Sabkha

Barth and Böer (2002a, b) enlisted the following characteristics of sabkhat: (1) flat areas with a crusty surface consisting of evaporite deposits, (2) periodically inundated either by seawater or precipitation, (3) found in tropical and sub-tropical arid areas with <250 mm/year precipitation, (4) shallow ground water level ranging between a few centimeters to two meters below surface and (5) mostly with poor plant cover except at the edges and areas with low salinity.

6.3 Research on Sabkha and Sabkha Ecosystems Book Series

Although many early explorers, such as Thomas (1929), Philby (1933), Harwood (1993) and Thesiger (1946) described various sabkhat and their characteristics in the early-mid eighteenth century; detailed scientific research based on geological and geographical aspects of sabkhat began in the 1960s followed by research on ecology and ecosystem services of sabkhat in the 1980s (Böer et al. 2019). For instance, the first comprehensive study of the origin and development of coastal Sabkha was conducted by Evans et al. (1969), who proposed a model of progradation and evaporation for the Abu Dhabi Sabkha. Gerdes et al. (1985) studied fauna of the Gavish Sabkha and the Solar Lake (along the Sinai coast, Egypt). Similarly, the first comprehensive exploration of the biota and ecology of Sabkha was done by Sheppard et al. (1992), who documented the diversity and distribution of microorganisms, plants and animals in the Abu Dhabi Sabkha. Since then, the number of research articles on sabkhat has been increasing progressively with time, linked to diverse science-disciplines, such as biology, conservation, ecology, geology, geography, geomorphology, and hydrology. In this context, the global book series 'Sabkha Ecosystems' with a production period exceeding 17 years (2002–2019),

Fig. 6.2 Sabkha Ecosystem book series (2002–2019) initially published through Kluwer Publishers and later by Springer-Nature

with a total of 148 chapters in six volumes (Barth and Böer 2002b; Khan et al. 2006, 2014, 2016; Öztürk et al. 2011; Gul et al. 2019), is considered the most comprehensive compilation of scientific documentation on sabkha ecosystems (Fig. 6.2). This book series provides a definition and classification of sabkhat, generated comprehensive scientific information on sabkha research, conservation and development from different parts of the world, and most importantly, it highlighted the knowledge-gaps, and suggested necessary future research. The first volume of the book series, which was published in 2002, covered the sabkhat of the Arabian Peninsula and adjacent countries. The second volume focused on research on the sabkhat in West and Central Asia. The third volume was on sabkhat of Africa and Southern Europe. The fourth volume dealt with cash crop halophytes and biodiversity conservation. The fifth volume covered research on the sabkhat of the Americas, while the sixth and last volume of the book series, published in 2019, focused on halophytes and sabkha ecosystems in the Asia-Pacific region. In general, this book series aimed to contribute to achieving the United Nations Sustainable Development Goals Zero Hunger, Clean Water, Clean Energy, Climate Action, Economic Growth, Sustainable Communities and Biodiversity on land and below water. It discussed important adaptation opportunities for climate change and clean energy, food security, and biodiversity conservation in the context of research related to sabkhat. It also outlined targets for achieving these goals using halophytes and sabkhat as resources. Despite an increase in research, sabkhat ecosystems require significantly more scientific research and documentation. Furthermore, the ecological importance of the sabkha has not received due attention (Böer et al. 2019). Still, little is known about the primary productivity, as well as the abiotic and biotic ecosystem components of sabkha ecosystems (Roda and Bassam 2017; Böer et al. 2019; Al-Dubai et al. 2022).

6.4 Plant and Algal Diversity of Sabkha Ecosystems

Sabkha ecosystems are unique and diverse, hosting a limited number of specially adapted phanerogames, algae, and blue-green algae, which can tolerate high salinity levels and harsh environmental conditions. Omar et al. (2002) reported that the sabkhat of Kuwait, which cover 10% of the country, are characterized by five major plant communities, i.e. *Nitraria retusa*, *Tamarix aucheriana*, *Zygophyllum qatarense*, *Halocnemum strobilaceum* and *Seidlitzia rosmarinus*. About 15 halophytes are found in the coastal and inland sabkhat of Oman (Ghazanfar 2002). Costa and Herrera (2016) reported 10 halophyte species from coastal flats/sabkhat and 16 halophytes from inland salt flats of Brazil. Khan and Gul (2002) reported that 108 halophyte species belonging to 36 families are found in the subtropical sabkhat of Pakistan and common species are *Suaeda fruticosa*, *Arthrocnemum macrostachyum*, *Cressa cretica*, *Salsola imbricata* and *Zygophyllum simplex*. Similarly, Ghazanfar et al. (2019) reported 120 taxa of halophytes mainly from Amaranthaceae, Zygophyllaceae, Plumbaginaceae, Poaceae, Cyperaceae, and Juncaceae. Some of the key halophytes of Arabia are *Avicennia marina* (mangrove), *Halocnemum strobilaceum* (glasswort), *Halopeplis perfoliata* (saltwort), *Arthrocnemum macrostachyum* (jointed glasswort), *Suaeda vermiculata* (seablite), *Limonium axillare* (sea lavender), *Sporobolus spicatus* (salt grass), and *Aeluropus lagopoides* (salt meadow grass). These plants have developed a number of specialized strategies to cope with high substrate salinity, such as salt secretion, succulence, osmotic adjustment, and salt-resistant seed germination (Ghazanfar et al. 2019).

Since sabkhat are periodically inundated, these ecosystems also support algal mats that are composed of cyanobacteria, diatoms, green algae, and red algae. These mats play an important role in stabilizing the sabkha surface, trapping sediments, and cycling nutrients. John (2002) reported the occurrence of *Enteromorpha*, *Chaetomorpha* and *Rhizoclonium* in the sabkhat; whereas he also reported that some others like *Hormophysa* and *Sargassum* can also be blown into sabkhat with high tides and storm surges. Likewise, Round (2002) reported diatoms belonging to *Mastogloia* and *Amorpha* in the Abu Dhabi sabkha. The diversity of plants and algae in sabkhat is influenced by spatial and environmental factors, such as elevation, soil texture, salinity, moisture, nutrients, temperature, wind, and human activities. A recent study by Al-Amro and Al-Qahtani (2022) found that the variation of the plant species composition in a sabkha of Saudi Arabia was related to environmental and spatial variables.

6.5 Animal Diversity of Sabkhat

According to a UNESCO World Heritage Centre document (https://whc.unesco.org/en/tentativelists/6352/), sabkhat are home to various animals, such as birds, reptiles, mammals, fish, crustaceans, mollusks, and insects. Some of the key animals of sabkhat include birds, such as flamingos, herons, egrets, gulls, terns, plovers, sandpipers, snipes, stilts, avocets, ospreys, falcons, eagles, owls, larks, warblers, shrikes, sparrows, as well as mammals, such as gazelles, foxes, hares, gerbils, jerboas, spiny mice, reptiles, such as lizards, snakes, turtles, crocodiles, and fish, curstaceans and shellfish, such as mudskippers, gobies, shrimps, crabs, clams, oysters, as well as snails, worms, and beetles. Loughland and Cunningham (2002) reported 10 mammal and 21 reptile species from the sabkhat of the Arabian Peninsula and 17 mammals and 9 reptiles from Central Asia. Similarly, Hogarth et al. (2002) reported the occurrence of insects belonging to the families of the Hemiptera, Homoptera, Lepidoptera, and Coleoptera from the sabkhat. These animals have developed various strategies to cope with salinity and temperature fluctuations, such as osmoregulation, thermoregulation, camouflage, burrowing, migration and nesting. The diversity of animals in sabkhat is influenced by spatial and environmental factors such as elevation, soil texture, salinity, moisture, nutrients, temperature, wind, and human activities. In addition, the diversity of animals in sabkhat may also be affected by the presence of algal mats and plant diversity that provide food and shelter for many organisms (Hogarth et al. 2002; Chenchouni et al. 2015).

6.6 Ecological Roles and Services of Sabkha Ecosystems

Sabkhat can provide valuable ecological roles and services, which can be summarized as follows:

1. ***Climate change mitigation***: Sabkhat can contribute towards mitigation of climate change by sequestering atmospheric CO_2 into their sediments (Elsayed et al. 2022). Sabkhat can also enhance the resilience of adjacent blue carbon ecosystems by buffering them from erosion, and salinity fluctuations (Barth and Böer 2002b).
2. ***Biodiversity conservation***: Sabkhat can conserve biodiversity by hosting unique and diverse communities of microorganisms, plants, and animals that are adapted to extreme conditions (Ghazanfar 2002; John 2002; Khan and Gul 2002; Loughland and Cunningham 2002). In addition, sabkhat can provide habitat connectivity and refuge to migratory and endangered species (Barth and Böer 2002b).
3. ***Coastal protection***: Sabkhat are important barriers for the protection of the coasts from erosion and flooding thereby stabilizing shorelines and dissipating wave energy (Loughland et al. 2016). Sabkhat can also minimize the impacts of storm surges and sea level rise by increasing the elevation and accretion of sediments (Loughland et al. 2016; Böer et al. 2019).
4. ***Fisheries and tourism support***: Sabkhat can support fisheries by providing nursery grounds and food sources for various fish species (Zajonz et al. 2002; Khan and Gul 2002) thereby supporting fisheries. Sabkhat can also offer opportunities for tourism and recreation based on their scenic beauty and cultural heritage.

6.7 Sabkhat as Blue Carbon Reserves

Blue carbon refers to the organic carbon that is captured and stored by coastal and marine ecosystems, such as mangroves, seagrasses, salt marshes, and algal mats. The carbon storage potential of sabkha ecosystems is not well studied and understood, but some studies have suggested that they may play a significant role in the global carbon cycle. Sabkhat are proposed to be capable of sequestering blue carbon, as they can contain significant amounts of organic carbon in their sediments. According to a study by Elsayed et al. (2022), sabkhat can store organic carbon ranging from 13.75 to 37.17 Mg C ha^{-1} in the top 30 cm of sediments, and from 31.25 to 84.48 Mg C ha^{-1} in the top 1 m of sediments. Generally, these values fall within the range of those reported for other blue carbon ecosystems, such as mangroves and salt marshes. Besides vegetation, another possible mechanism of carbon storage in sabkhat is the formation of carbonate minerals, such as dolomite and calcite, through microbial and geochemical processes. Carbonate minerals can sequester carbon for long periods of time and preserve isotopic signatures of past environmental conditions (Brasier et al. 2011). Therefore, sabkhat, especially those in coastal areas, can be considered as blue carbon ecosystems, in *sensu lato*, depending on the presence and extent of halophyte vegetation and microbial activity. Moreover, sabkhat can also contribute to carbon storage by providing habitat and substrate for blue carbon plants and microorganisms as well as by preventing or reducing the degradation or loss of adjacent blue carbon ecosystems.

6.8 Threats to Sabkha Ecosystems

Sabkha ecosystems are valuable for their ecological functions such as carbon sequestration, erosion control, habitat provision, and phytoremediation. However, in the past, sabkhat were believed to be wastelands and thus many of these ecosystems have undergone rapid and 'uncoordinated' development, caused by landfilling, top-soiling, dredging, and construction (Fig. 6.3; Shaer 2009, 2010; Ghazanfar et al. 2019; Guerine et al. 2020). Lokier (2013) enlisted as many as 13 anthropogenic threats to the sabkha environment, which can have various degrees of physical, geochemical, and hydrological impacts on sabkha. In addition, sabkhat are also threatened by various other factors such as overgrazing, oil pollution, invasive species and climate change (Al-Jeneid et al. 2008; Naser 2014). Climate change driven sea-level rise is becoming a serious threat to the Sabkha ecosystems of the world (Hassan and Hassaan 2020; Al-Jeneid et al. 2008; Babu et al. 2012). Sea level rise will increase the extent and frequency of inundation of Sabkha soils, which can in turn alter hydrology, salinity, sedimentation, and biogeochemistry of sabkhat, thereby influencing their overall physical and biological characteristics. Sea level rise can also increase the wave action and coastal erosion of sabkhat, especially along their seaward margins. The impacts of sea level rise on sabkha ecosystems can

Fig. 6.3 Coastal Sabkha along Hawksbay coast Karachi (Pakistan) has been subjected to very intense construction activities

have substantial socio-economic implications for the human populations that depend on them for their livelihoods and well-being. For example, sea level rise can affect the productivity and sustainability of salt production, fisheries, agriculture, tourism, and recreation in sabkha areas. There is a need for science-based landscape planning, and to conserve and restore these ecosystems, so they can maintain their biodiversity and ecosystem services.

6.9 Conservation and Management of Sabkhat

As detailed above, sabkhat, despite being ecologically and economically important, are under threat, and therefore need urgent attention towards conservation and management. The following measures could be important for the conservation and management of Sabkhat:

(i) ***Periodic assessments***: Periodic assessments/evaluations of the type, extent, trends, and drivers of the threats to sabkha is necessary to monitor and understand the current situation and future scenarios, which can provide a scientific basis for decision-making and policy-development (Lokier 2012; Chenchouni et al. 2015; Rajendran et al. 2021; Abyat et al. 2021).

(ii) ***Conservation/restoration efforts***: Well-planned conservation and restoration efforts based on scientific assessments are necessary to maintain or even enhance functions and services of Sabkha ecosystems, including their immensely important role in geo-heritage conservation, and in geo-tourism. These efforts may include: establishing protected areas, including UNESCO-designated sites, such as Biosphere Reserves, Global Geoparks, World Natural Heritage Sites, reducing disturbances, controlling habitat degradation, controlling invasive species, encouraging sustainable use, protecting biodiversity, enhancing soil stability, maintaining the hydrological balance, and arranging for educational and awareness campaigns, including field-excursions (Clüsener-Godt 2002; Kleo and Al-Otaibi 2011; Lokier 2013).

(iii) ***Socio-economic accounting***: It is imperative to evaluate the economic, socio-cultural, heritage, educational, and ecological benefits of sabkhat in order to highlight their significance and support their preservation/conservation (Krumbein 1985; Clüsener-Godt 2002; Salem et al. 2019). In order to do this, it may be necessary to estimate the monetary or non-monetary value of the sabkha ecosystem services, create incentives or payments for ecosystem services, and create awareness amongst local communities about benefits of sustainable utilization of Sabkha resources.

(iv) ***Scientific research***: It is of the utmost importance, in terms of anthropogenically accelerated climate change and sea-level fluctuations, to carry out significantly more scientific research, including M.Sc. and Ph.D. thesis, into the carbon sequestration capacity, ecology, chemistry, geology, geography, geomorphology, and ecosystem succession of sabkha ecosystems, and adjacent ecosystems, including blue-green algal mats, salt marshes, mangrove forests, and seagrass beds, under sea-level fluctuation. With sea-level rise we can expect sabkhat to gradually convert into salt masrhes, then mangroves, and then seagrass beds, depending on the intensity of inundation. And, vice versa, we could also expect mangroves, salt marshes, blue-green algal mats, and seagrass beds to gradually convert into sabkhat, if the sea-level would decline again.

6.10 Conclusions

Sabkha is a salt flat found in arid or semi-arid regions, naturally generated due to the evaporation of saline water. In recent decades, sabkhat were considered wastelands and received little attention from research, conservation and socio-economic viewpoints. However, research over the last four decades has not only unveiled the origin, as well as geochemical, mineralogical, microbiological and paleontological aspects, of sabkha, but also highlighted the ecological importance of Sabkha ecosystems. It is also now appreciated that the sabkhat provide various ecosystem services such as carbon sequestration, nutrient cycling, wildlife habitat for numerous vertebrate and invertebrate species, and shoreline protection alongside different economic benefits (e.g. salt production and geo-tourism). However, sabkhat are threatened by human activities such as urbanization, industrialization, land reclamation, groundwater extraction, and pollution. Hence, there is a need for conservation and management of these important habitats and heritage landscapes.

References

Abyat K, Landi A, Amerikhah H (2021) Evaluation of sabkha changes caused by Houralimiz wetland during 20 years using Landsat images. J Water Soil Conserv 28(1):165–181

Al-Amro AM, Al-Qahtani SM (2022) Plant diversity in Sabkha ecosystems of arid region: spatial and environmental drivers. Braz J Biol 82:e262331

Al-Dubai TA, Bantan RA, Abu-Zied RH, Al-Zubieri AG, Jones BG (2022) s. Front Mar Sci 9:351

Al-Jeneid S, Bahnassy M, Nasr S et al (2008) Vulnerability assessment and adaptation to the impacts of sea level rise on the Kingdom of Bahrain. Mitig Adapt Strat Glob Change 13:87–104. https://doi.org/10.1007/s11027-007-9083-8

Babu DSS, Sivalingam S, Machado T (2012) Need for adaptation strategy against global sea level rise: an example from Saudi coast of Arabian gulf. Mitig Adapt Strateg Glob Change 17:821–836. https://doi.org/10.1007/s11027-011-9346-2

Barth HJ (2001) Comment on "Playa, playa lake, sabkha: proposed definitions for old terms". J Arid Environ 4(47):513–514

Barth H-J, Böer B (2002a) Introduction. In: H-J Barth, B Böer (eds) Sabkha Ecosytems – volume I (The Arabian Peninsula and adjacent countries), Kluwer Academic Publishers, Dordrecht. ISBN978-90-481-5972-7, pp 1–5

Barth H-J, Böer B (2002b) Sabkha Ecosytems – Volume I (The Arabian Peninsula and Adjacent Countries). Kluwer Academic Publishers, Dordrecht

Böer B, Clüsener-Godt M, Gul B, Khan MA, Hameed A (2019) Sabkha ecosystems Vol VI: Asia-Pacific—summarizing the story. In: Gul B, Böer B, Khan M, Clüsener-Godt M, Hameed A (eds) Sabkha ecosystems, Tasks for vegetation science, vol 49. Springer, Cham. https://doi.org/10.1007/978-3-030-04417-6_32

Brasier AT, Fallick AE, Prave AR, Melezhik VA, Lepland A, Scientists FAR-DEEP (2011) Coastal sabkha dolomites and calcitised sulphates preserving the Lomagundi-Jatuli carbon isotope signal. Precambrian Res 189(1–2):193–211. https://doi.org/10.1016/j.precamres.2011.05.011

Briere PR (2000) Playa, playa lake, sabkha: proposed definitions for old terms. J Arid Environ 45(1):1–7

Chenchouni H, Menasria T, Neffar S, Chafaa S, Bradai L, Chaibi R, Mekahlia MN, Bendjoudi D, Si BA (2015) Spatiotemporal diversity, structure and trophic guilds of insect assemblages in a semi-arid Sabkha ecosystem. PeerJ 3:e860. https://doi.org/10.7717/peerj.860

Clüsener-Godt M (2002) UNESCO's work for the protection of arid zones with particular emphasis to sabkhat. Sabkha Ecosyst 1:347–352

Costa CSB, Herrera OB (2016) Halophytic life in Brazilian salt flats: biodiversity, uses and threats. Sabkha Ecosyst: volume V. The Americas, pp 11–27

Elsayed H, Al Disi Z A S, Naja K, Strakhov I, A Rahman S A Al-kuwari H, Sadooni F, Al-Khayat J A A, Dittirich M (2022) Assessment of blue carbon retention in siliciclastic and carbonate-dominated sabkhas ecosystems. Available at SSRN: https://ssrn.com/abstract=4306118 or https://doi.org/10.2139/ssrn.4306118

Evans G, Schmidt V, Bush P, Nelson H (1969) Stratigraphy and geologic history of the sabkha, Abu Dhabi, Persian Gulf. Sedimentology 12:145–159

Gerdes G, Spira J, Dimentman C (1985) The fauna of the Gavish Sabkha and the solar Lake—a comparative study. In: Hypersaline ecosystems: the Gavish Sabkha. Springer, Berlin, Heidelberg, pp 322–345

Ghazanfar SA (2002) The sabkha vegetation of Oman. In: Barth H-J, Böer B (eds) Sabkha ecosytems – volume I (The Arabian Peninsula and Adjacent Countries). Kluwer Academic Publishers, Dordrecht. ISBN978-90-481-5972-7, pp 99–108

Ghazanfar SA, Böer B, Al Khulaidi A W, El-Keblawy A, Alateeqi S (2019) Plants of Sabkha ecosystems of the Arabian peninsula. Sabkha ecosystems: volume VI: Asia/Pacific, 55–80

Glennie KW (1970) Desert sedimentary environments, Developments in sedimentology, 14. Elsevier, p 222

Goodall TM, Al-Belushi JD (2020) A glossary of Arabic desert terminology used in southeastern Arabia. In: Quaternary deserts and climatic change. CRC Press, pp 611–619

Guerine L, Belgourari M, Guerinik H (2020) Cartography and diachronic study of the Naama sabkha (Southwestern Algeria) remotely sensed vegetation index and soil properties. J Rangel Sci 10(2):172–187

Gul B, Böer B, Khan MA, Clüsener-Godt M, Hameed A (2019) Sabkha ecosystems – volume VI, Asia/Pacific. Springer

Harwood GM (1993) Sabkha. In: Keary P (ed) The encyclopedia of the solid earth sciences. Blackwell Science Ltd, Oxford, pp 532–534

Hassan A, Hassaan MA (2020) Potential impact of sea level rise on the geomorphology of Kuwait state coastline. Arab J Geosci 13:1139. https://doi.org/10.1007/s12517-020-06084-1

Hogarth PJ, Tigar BJ, Barth HJ, Böer B (2002) Ecology of sabkha arthropods. In: Barth H-J, Böer B (eds) Sabkha Ecosytems – volume I, The Arabian peninsula and adjacent countries. Kluwer Academic Publishers, Dordrecht. ISBN978-90-481-5972-7, pp 267–282

John DM (2002) Macroalgae associated with sabkha. In: Barth H-J, Böer B (eds) Sabkha ecosytems – volume I, The Arabian Peninsula and Adjacent Countries. Kluwer Academic Publishers, Dordrecht. ISBN978-90-481-5972-7, pp 239–246

Khan MA, Gul B (2002) Salt tolerant plants of coastal Sabkhat of or Pakistan. In: Barth H-J, Böer B (eds) Sabkha Ecosytems – volume I, The Arabian peninsula and adjacent countries. Kluwer Academic Publishers, Dordrecht. ISBN978-90-481-5972-7, pp 123–140

Khan MA, Böer B, Kust GS, Barth HJ (2006) Sabkha ecosystems – volume II. (West and Central Asia), Springer
Khan MA, Böer B, Öztürk M, Al Abdessalaam TZ, Clüsener-Godt M, Gul B (2014) Sabkha ecosystems – volume IV. (Cash Crop Halophyte and Biodiversity Conservation), Springer
Khan MA, Boër B, O̊zturk M, Clüsener-Godt M, Gul B, Breckle SW (2016) Sabkha ecosystems – volume V. (The Americas). Springer
Kinsman DJ (1969) Modes of formation, sedimentary associations, and diagnostic features of shallow-water and supratidal evaporites. AAPG Bull 53(4):830–840
Kleo AA, Al-Otaibi O (2011) The sustainable development of Kuwaiti Sabkhas. Dig Middle East Stud 20(1):27–49
Krumbein WE (1985) Applied and economic aspects of sabkha systems—genesis of salt, ore and hydrocarbon deposits and biotechnology. Hypersaline Ecosystems: The Gavish Sabkha, 426–436
Lokier SW (2012) Development and evolution of subaerial halite crust morphologies in a coastal sabkha setting. J Arid Environ 79:32–47
Lokier SW (2013) Coastal Sabkha preservation in the Arabian Gulf. Geoheritage 5:11–22. https://doi.org/10.1007/s12371-012-0069-x
Loughland RA, Cunningham PL (2002) Vertebrate fauna of Sabkhat from the Arabian peninsula: a review of Mammalia, Reptilia and Amphibia. In: Barth H-J, Böer B (eds) Sabkha Ecosytems – volume I, The Arabian peninsula and adjacent countries. Kluwer Academic Publishers, Dordrecht. ISBN978-90-481-5972-7, pp 255–266
Loughland RA, Qasem AM, Burwell B, Prihartato PK (2016) Coastal Sabkha (Salt flats) of the southern and Western Arabian Gulf. In: Finlayson C, Milton G, Prentice R, Davidson N (eds) In encyclopedia of wetlands: wetlands of the world, Coastal Sabkha. Ecosystems book, Springer, The wetland book, vol 4. Springer, Dordrecht. https://doi.org/10.1007/978-94-007-6173-5_185-3
Naser HA (2014) Marine ecosystem diversity in the Arabian Gulf: threats and conservation. Biodiversity–The Dynamic Balance of the Planet, 297–328
Neal JT (1975) Playas and Dried Lakes. Dowden, Hutchinson & Ross. Inc., Stroudsberg
Omar SAS, Misak RF, Shahid SA (2002) Sabkha and halophytes in Kuwait. In: Barth H-J, Böer B (eds) Sabkha Ecosytems – volume I (The Arabian peninsula and adjacent countries). Kluwer Academic Publishers, Dordrecht. ISBN978-90-481-5972-7, pp 71–82
Öztürk M, Böer B, Barth HJ, Breckle SW, Clüsener-Godt M, Khan MA (2011) Sabkha ecosystems – Volume III (Africa and Southern Europe). Springer
Philby HSJ (1933) Rub'Al Khali: an account of exploration in the great South Desert of Arabia under the auspices and patronage of his Majesty'Abdul'Aziz ibn Sa'ud, King of the Hejaz and Nejd and its dependencies. Geogr J 81:1–21
Rajendran S, Al-Kuwari HAS, Sadooni FN, Nasir S, Govil H (2021) Remote sensing of inland Sabkha and a study of the salinity and temporal stability for sustainable development: a case study from the west coast of Qatar. Sci Total Environ 782:146932
Roda FAT, Bassam TY (2017) Halo-thermophilic bacteria and heterocyst cyanobacteria found adjacent to halophytes at Sabkhas, Qatar: preliminary study and possible roles. Afr J Microbiol Res 11(34):1346–1354
Round FE (2002) Diatoms of the sabkha. In: Barth H-J, Böer B (eds) Sabkha Ecosytems – volume I (The Arabian peninsula and adjacent countries). Kluwer Academic Publishers, Dordrecht. ISBN978-90-481-5972-7, pp 247–254
Salem S, Nasri S, Abidi S, Smaoui A, Nasri N, Mutjé P, Ben Hamed K (2019) Lignocellulosic biomass from Sabkha native vegetation: a new potential source for fiber-based bioenergy and bio-materials. In: Sabkha ecosystems: volume VI: Asia/Pacific, pp 407–412
Shaer HME (2009) Potential role of Sabkhas in Egypt: an overview. In: Ashraf M, Ozturk M, Athar H (eds) Salinity and water stress: improving crop efficiency. Springer, Dordrecht, pp 221–228
Shaer HME (2010) Potential role of salt marshes in the Sabkhas of Egypt. In: Sabkha ecosystems: volume III: Africa and Southern Europe. Springer Netherlands, Dordrecht, pp 95–98

Sheppard CRC, Price P, Roberts C (1992) Marine ecology of the Arabian region. Academic, London, pp 1–347
Thesiger W (1946) A new journey in southern Arabia. Geogr J 108(4/6):129–145
Thomas B (1929) The south-eastern borderlands of Rub'al Khali. Geogr J 73(3):193–212
Whitten DGA, Brooks JRV (1972) Penguin dictionary of geology. Penguin Books Ltd.
Zajonz U, Beech MARK, Gill AC (2002) Fishes of Sabkha-related habitats. In: Barth H-J, Böer B (eds) Sabkha ecosystems, Dordrecht, pp 283–298

Chapter 7
Sustainable Development of Mangrove Ecosystems: A Blue Carbon Perspective

Daniel M. Alongi and Martin Zimmer

Abstract Mangroves sequester more Blue Carbon (BC) than most other marine ecosystems, making them ideal candidates for conservation and sustainable (re-) establishment, which is urgently needed due to large global losses. Deforestation rates are declining, but conservation of remaining intact forests and/or forests of high intrinsic value are still urgently needed. Conserving existing mangroves is the most cost-effective and efficient measure for management of C sequestration. Forest losses result in oxidation of sediment organic matter and the return of large amounts of BC as CO_{2eq} to the atmosphere. Losses from sea-level rise are forecast to increase by 1.1–3.4 Pg CO_{2eq} by 2100. Model projections of deforestation rates into the future suggest emissions of 679 Tg CO_{2eq} from 2012–2095.

Global CO_{2eq} emissions from all causes (mostly erosion, agriculture, aquaculture) may reach 2391–3392 Tg CO_{2eq} by 2100, with largest losses from the West Coral Triangle. Sustainable development of mangrove restoration/rehabilitation or afforestation projects is still in its infancy with environmental, technical, social, economic, and political problems and barriers; most mangrove BC projects are unsuccessful. Success rates can be enhanced by: improving policy and legal arrangements to ensure equitable benefit sharing; improving stewardship by incorporating knowledge and values of local communities; clarifying property rights; improving financial approaches and accounting tools to incorporate co- benefits; developing low cost, technological methods to measure carbon sequestration; and resolving knowledge gaps in BC biogeochemistry. Mangroves buffer only less than 1% of global CO_2 emissions, but restoration projects can be of local and national importance.

D. M. Alongi (✉)
Tropical Coastal & Mangrove Consultants, Pakenham, VIC, Australia
e-mail: dmalongi@outlook.com

M. Zimmer
Leibniz Centre for Tropical Marine Research (ZMT) GmbH, Bremen, Germany
e-mail: martin.zimmer@leibniz-zmt.de

M. Clüsener-Godt et al. (eds.), *Blue Carbon Mangrove Ecosystems*, Blue Carbon Ecosystems for Sustainable Development 1,
https://doi.org/10.1007/978-3-031-69553-7_7

Keyword Blue carbon · Carbon · Conservation · Mangroves · Restoration · Sustainable development

7.1 Introduction

Mangrove forests, salt marshes, and seagrass meadows are recognized as coastal blue carbon ecosystems (BCEs) because they facilitate the sequestration of carbon (C), mostly in their sediments, playing key roles in C cycling in the coastal ocean (Alongi 2018).

Mangrove forests are the world's most C-rich ecosystems to the extent that they help mitigate anthropogenic CO_2 emissions (Alongi 2022). Despite their ecological and economic importance, mangroves are still being destroyed (Goldberg et al. 2020), although there has been a clear reduction in rates of forest losses (Friess et al. 2020), necessitating conservation where still possible and re-establishment where necessary and feasible. Further, mangroves are exposed to land-use change and climate change, along with all their consequences, such as sea-level rise, erosion, and habitat loss due to extreme weather events. When these forests are degraded or destroyed, organic matter (OM) in their sediments is oxygenated and decays, resulting in the release of CO2 to the atmosphere (Lovelock et al. 2011) and as porewater DIC released via groundwater (Alongi 2020a).

The UN Sustainable Development Goals (SDGs) provide several opportunities to prioritise conservation and re-establishment of ecosystems as well as to explicitly recognize the role of coastal wetlands in ameliorating climate change. SDG 14 specifically considers coastal and marine environments to conserve and sustainably use their resources for sustainable development (https://sdgs.un.org/goals/goal14). Mangroves contribute many ecosystem services to SDG 14, including seafood production for fisheries, coastal protection, acidification regulation, and C sequestration and storage.

Mangrove forests are leading candidates for several market-based approaches for forest conservation, including payments for ecosystem services (PES) and C finance marketing schemes to address some of the underlying drivers of forest loss and to provide direct economic incentives for conservation. Mangroves are particularly well suited to generate C credits because of their unrivalled ability to store C at high rates and in large amounts, their ecological resistance and resilience to disturbances, and their extensive provision of ecosystem services other than C sequestration ('co-benefits'). Thus, mangrove forests are 'the blue carbon wealth of nations' (Bertram et al. 2021), and by adopting a 'carbon centric approach' (Locatelli et al. 2014), multiple opportunities for blue economy are available to sustainably develop mangrove forests.

BC strategies must focus on maximizing forest production, including the sustainable use of wood resources, but primarily to facilitate the preservation and enhancement of organic carbon (C_{ORG}) stocks and burial capacity in mangrove sediment. Mangroves also accumulate C_{ORG} in plant biomass as the forests grow and age to maturity. Like in most other BCEs, the complex three-dimensional structure of mangrove aboveground biomass captures and deposits sediment particles and associated inorganic and organic C. While much tree C_{ORG} is eventually lost due to clear cutting (but see above for sustainable wood-use), or natural senescence and subsequent decay, and exported to adjacent coastal waters, 75–95% of mangrove C_{ORG} is stored belowground (Alongi 2014) to depths of several meters, especially in dead roots and rhizomes and as recalcitrant OM of multiple autochthonous and allochthonous origins (Zimmer and Helfer 2020). Under the right conditions, C_{ORG} is stored as peat (Alongi 2018). Peat formation is a function not only of OM inputs, tree species, litter composition, microbial and faunal activities, but also moisture and temperature, slow decay rates of refractory material, and the magnitude and frequency of tides. C_{ORG} in mangrove deposits may be stored for hundreds or even thousands of years, facilitated by several unique physical, chemical, and hydrological drivers (Cohen et al. 2012; Collins et al. 2021). C_{ORG} accumulation generally depends on several factors such as tidal amplitude, forest elevation, location in relation to the open coast and in relation to a tidal waterway, tree density, and species composition. Mangroves capture silt, clay, and organic particles being maintained in suspension in turbulent wakes created by the movement of tidal waters around tree trunks, prop roots and pneumatophores, and fallen timber (Le Minor et al. 2019). Most particles and organic flocs settle around slack tide with most being retained within the forest despite the pull of ebb tide. The sticking microbial mucus on particle surfaces and the formation of egested fecal pellets by invertebrates facilitate rapid settling of particles (Alongi 2009).

To develop sustainable management frameworks and to understand the benefits and costs of mangrove re-)establishment (*sensu* Zimmer et al. 2022), BC resources need to be fully integrated into climate change mitigation and conservation strategies on local, national, and global scales (Hilmi et al. 2021). This chapter focuses on the sustainable development of mangrove forests through the lens of BC biogeochemistry and ecology, especially CO_2 and CH_4 emissions and BC stock changes due to various disturbances and disturbance-related processes. Advances can then be made to unleash their potential for mitigating climate change and to maintain and conserve their other ecosystem services.

7.2 Blue Carbon Biogeochemistry

7.2.1 An Ecosystem-Level View

The most sophisticated estimates of global values to date still rely on global averages that are subject to considerable uncertainty, as they are based on upscaling of highly variable local and regional values. Future research activities should aim at closing the geographical gaps in coverage of accurate measurements of C storage rates and stocks in mangroves (and other BCEs).

A mass balance of C flow through an intact, undisturbed mangrove ecosystem (Table 7.1) based on global means (Alongi 2020a) shows that about 63% of gross

Table 7.1 Mass balance of C in an idealized, undisturbed mangrove forest

Inputs		Outputs	
GPP (mangrove)	35.3	R_{CANOPY}	22.3
GPP (microalgae)	4.4	$R_{MICROALGAE}$	2.3
DIC production[a]	18.4	$R_{TIDAL\ WATER}$	3.35
Marine	NA	Air-sediment CH_4 release	0.015
Catchment	NA	Air-water CH_4 release	0.0116
Groundwater	NA	Sediment CO_2 + DIC release (air + water)	6.13
		Net DIC export	14.0
		Net DOC export	5.90
		Net POC export	1.73
		Net CH4 export	0.0277
		Burial	1.11
Total	**57.9**	**Total**	**56.9**

(Mg C ha^{-1} a^{-1}) based on globally averaged data (Alongi 2020a). *GPP* gross primary production, R_{CANOPY} forest canopy respiration, $R_{MICROALGAE}$ microalgal respiration, $R_{TIDAL\ WATER}$ tidal water respiration, *DIC* dissolved inorganic carbon, *DOC* dissolved organic carbon, *POC* particulate organic carbon, CH_4 methane, *NA* no data
[a]sediment depth of 1 m

primary production (GPP) is returned to the atmosphere as canopy respiration. Like tropical terrestrial forests (Malhi et al. 2011), nearly equal amounts of net primary productivity are vested in wood (3.7 Mg C ha^{-1} a^{-1}), canopy (4.3 Mg C ha^{-1} a^{-1}) and root (5.2 Mg C ha^{-1} a^{-1}) production, indicating patterns of C allocation nearly identical to their terrestrial counterparts. The C cycle is in overall balance (inputs minus outputs) in undisturbed mangroves, but this is unlikely to hold true for disturbed forests. For example, mangrove stands across Indonesia, with contrasting environmental settings and conditions of disturbance, exhibit large differences in both C storage rates and stocks, differing from each other by factors of almost 9 and almost 4, respectively (Kusumaningtyas et al. 2019, 2022). Further, there are huge variations across different regions, mangrove forests, and even within one forest at rather small spatial scales. For instance, within a single area, such as in northern Brazil (Quadros et al. 2019), litter production values range from 0.4 to 10.4 Mg ha^{-1} a^{-1} for *R. mangle,* 0.5–6.2 Mg ha^{-1} a^{-1} for *A. germinans* and 0.04–0.6 Mg ha^{-1} a^{-1} for *L. racemosa.* C burial rates in mangrove forests range globally from 1.5 to 29 kg C ha^{-1} d^{-1} (Rosentreter et al. 2021; Hatje et al. 2021). Litter-based carbon sinks in BCEs will likely increase upon global warming (Ouyang et al. 2023).

Despite limitations to the accuracy of global estimates, the C mass balance identifies the main inputs as mangrove forest primary production and net dissolved C production in mangrove sediments to at least a depth of 1 m (Table 7.1). Inputs from terrestrial and marine sources have been identified and can be significant but are highly variable and site-specific (see below). Net DIC + DOC production is not normally considered as an input in C mass balance models, but evidence indicates that centuries-old, sequestered OM, likely derived from pre-mangrove vegetation or earlier deposition in a pre-mangrove habitat (e.g., mudflat), remains susceptible to remineralization and tidal DIC and DOC export (Maher et al. 2017), if exposed to air or oxidizing conditions. Radiocarbon dating suggests that mangrove sediments in Brazil, India and China over a 1–3 m depth profile are centuries-old, certainly older than the existing mangrove vegetation (Passos et al. 2021; Rani et al. 2021; Zhang et al. 2021). This implies a deep reservoir of utilizable C that originates from sources prior to modern mangrove colonization, such as Holocene terrestrial, freshwater, estuarine and/or marine OM deposits, including ancient mangrove peat.

Further, C inputs via groundwater, surface runoff, and inputs of allochthonous C_{ORG} from the sea, land, and catchment likely play an important role in mangrove C dynamics, considering that about 42% of mangrove sediment OM is derived from external sources (Kristensen et al. 2008, 2017). Several studies measuring sedimentary C:N ratios and $\delta^{13}C$ profiles (Chen et al. 2018; Elmstrom et al. 2018; Xiong et al. 2018; Hargan et al. 2020; Matos et al. 2020; Minu et al. 2020; Sasmito et al. 2020; Belliard et al. 2022; Ray et al. 2022; Suello et al. 2022) indicate allochthonous C contributions to mangrove sediments ranging from undetectable to mostly marine. In restored mangroves of various ages in China, the proportion of sediment OM derived from allochthonous suspended matter increased with increasing forest age (Chen et al. 2018). Also in China, Xiong et al. (2018) found that mangrove C_{ORG} accounted from 27–97% of mangrove sediment C, varying with intertidal position, site location, and forest age. Similarly, changes in the ratio of terrestrial to marine inputs to mangrove forests varied along the Rufiji delta, Tanzania (Minu et al. 2020), with forests in the northern delta receiving mostly recycled OM and marine microbial detritus and central and southern delta forests receiving more anthropogenic and terrestrial inputs. In the Amazon, 49–95% of suspended OM is derived from marine phytoplankton in tidal creeks, implying significant marine inputs to mangrove forests lining the creek beds.

One of the most recent OM studies (Belliard et al. 2022) found increasing marine phytoplankton input moving downstream along the Guayas delta in Ecuador, with more terrestrial C inputs upstream. A similar gradient of changes in the origin of sediment OC was observed along the Aklan River estuary in the Philippines (Ray et al. 2022) and in the Guayas delta, Ecuador, where younger stands consist of more allochthonous C (79%) than older forests (59%), with an average of 65% of sediment OC derived from non- mangrove sources (Suello et al. 2022). Examining changes in sources and degradation of OM downcore in an urbanized mangrove forest in SW India, Cheriyan et al. (2022) measured a shift from more land-plant derived OM in deeper deposits to more marine algal OM in upper core sediments. Further, there was selective preservation of bacterial biomass down-core, with a transformation from labile to increasingly refractory OM with increasing sediment depth.

C outputs from mangrove forests are currently easier to identify than inputs (Table 7.1). After canopy respiration, export of DIC and DOC are the second major outputs as nearly all C respired within the entire sediment horizon is released in dissolved form via advective porewater exchange and/or lateral transport and subsurface tidal pumping to adjacent tidal waters (Table 7.1). Mangroves inhabit only 0.3% of the global coastal ocean area, but contribute 55% of air-sea CO_2 exchange, 14% of C_{ORG} burial, 28% of DIC export, and 13% of POC + DOC export from the world's coastal wetlands and estuaries to the atmosphere and global coastal ocean (Alongi 2020a). The fate of the OM exported to the sea is essentially unknown; depending on its chemical composition and structure, it can be decayed and respired microbially (adding to DIC) in the water column or remain stable as refractory DOM or POM and be deposited at the sea bottom over centuries.

7.2.2 *Blue Carbon Sequestration*

Rates of CORG burial in mangrove sediments average 220 g C_{ORG} m^{-2} a^{-1} (± 1SE = 18 g C_{ORG} m^{-2} a^{-1}) with a median of 128 g C_{ORG} m^{-2} a^{-1} over a very wide range from 0.001 to 1844 g C_{ORG} m^{-2} a^{-1} (c.f. Rosentreter et al. 2021, Alongi 2022). More than half of all measurements were < 200 g C_{ORG} m^{-2} a^{-1} (Fig. 7.1), with the highest rates observed in heavily impacted mangrove forests such as in China, South and Southeast Asia, and Africa. These high rates were likely caused by severe deforestation and subsequent sediment erosion upstream, resulting in massive bedload transport and deposition downstream into tidal flats and mangrove forests (Alongi 2018). In the Ciénaga Grande de Santa Marta, Caribbean Colombia (Perdomo Trujillo et al. 2021), sediment C_{ORG} stocks were twice as large in areas with high mangrove mortality as compared to conserved mangroves. The authors concluded that the

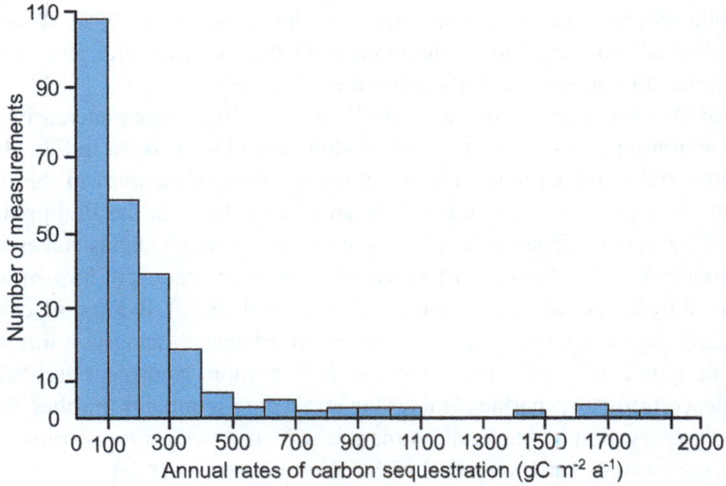

Fig. 7.1 Annual rates of blue carbon sequestration (g C m^{-2} a^{-1}) in sediments in various mangrove forests worldwide. Sources: Alongi (2018), Salmo III and Gianan (2019), Liu et al. (2020), Hanggara et al. (2021), Hatje et al. (2021), Kandasamy et al. (2021), Wigand et al. (2021), Carnell et al. (2022), Hamzeh and Lahijani (2022), Kusumaningtyas et al. (2019, 2022), Ratul et al. (2022), Rovai et al. (2022), and Zhang et al. (2022)

accumulation of dead plant material due to previous hyper-salinization of the embayment in concert with low decay rates in the anoxic sediment drove these large accumulation rates. Other mangroves expecting high mortality, such as from extreme weather events, may similarly show large amounts of carbon burial.

Multiplying the annual median sequestration rate (128 g C_{ORG}) by the most recent and accurate estimate of mangrove area (145,068 km^2; Jia et al. 2023), annual C burial in the world's mangrove forests equates to 21.1 Tg C_{ORG} which is several times greater than for annual burial in tropical coastal sediments (3.9 Tg C_{ORG}), and slightly greater than for salt marshes (11.7 Tg C_{ORG}) but less than in seagrass meadows (35.3 Tg C_{ORG}) (Alongi 2020b).

The wide variability in C_{ORG} burial rates reflects large differences in forests of different age, type, location, and disturbance history. Rates are a function of multiple interrelated drivers such as geomorphology, tidal inundation frequency, forest age, species composition, storms, tidal prism, sediment grain size, and oceanic, riverine, and anthropogenic inputs; sequestration rates are not significantly related to latitude (Mackenzie et al. 2021). In many tropical river deltas, sediment deposition and subsequent C_{ORG} accumulation and burial in mangrove sediments are often highly dynamic, with episodes of sediment starvation due to damming alternating with episodes of massive bedload transport and deposition that smother mangroves, as found in the Mekong delta (Collins et al. 2021).

In some forests, increased C_{ORG} sequestration rates with stand age have been measured, likely a result of increased tree density and thus more efficient sediment and particle trapping (Kathiresan et al. 2013; Soper et al. 2019; Dontis et al. 2020; Kandasamy et al. 2021), although sequestration likely stabilizes after maturity (Carnell et al. 2022). In southwest Florida, the restoration of sediment C sequestration potential shows a likely trajectory of increased rates to about a century after the initiation of restoring tidal connectivity (Cormier et al. 2022). However, C_{ORG} burial rates in newly restored and recently harvested mangroves are usually less than in protected, and thus older, mangrove forests (Salmo III and Gianan 2019; Hanggara et al. 2021; Kusumaningtyas et al. 2019, 2022).

7.3 Can Carbon Losses Be Managed?

Most mangrove forests worldwide are disturbed to some extent, from small-scale logging to total deforestation or degradation. Naturally, different scenarios result in different amounts of BC loss, either as exported dissolved or particulate C_{ORG} from sediments to adjacent waters, or lost as burnt biomass and thus CO_2 and CH_4 emissions to the atmosphere. Disturbance of mangrove sediments results in changes in oxygenation and redox state and pathways of microbial respiration, resulting in immense dissolved carbon (DC) losses via submarine groundwater discharge.

7.3.1 Deforestation and Climate Change

Modelling the effects of deforestation and climate change on past, present, and future CO_2 equivalents (CO_{2eq}) under two climate scenarios (RCP4.5 = "business as usual" and RCP8.5 = high emissions), Chatting et al. (2022) found that emissions from mangrove deforestation have declined from 141% to just above 6% of annual C sequestration rates since the 1980s (Fig. 7.2). Projecting current mangrove deforestation rates into the future will result in total emissions of 678 ± 151 Tg CO_{2eq} from 2012–2095. Further reductions in deforestation rates would elevate the compensation of C emissions from climate change by 55% under a "business as usual" scenario to 61% under a high emissions scenario, making the idea of offsetting emissions through mangrove conservation and (re-) establishment more viable.

Sea-level rise is also predicted to impact mangrove BC stocks and emissions, depending on emission scenarios and the availability of space for mangrove stands to migrate landwards ('coastal squeeze') (Lovelock and Reef 2020). Modelling indicates that C_{ORG} stocks and CO_{2eq} emissions are greatest under a worst-case scenario of high coastal squeeze and RCP8.5, with estimated losses of 3.4 Pg C and net gains of BC stocks of about 1.5 Pg by 2100. Conversely, lesser amounts of BC are forecast to be lost under low coastal squeeze and the RCP4.5 scenario, with predicted CO_{2eq} emissions of 1.1 Pg C and global gains and losses of about −0.3–0.5 Pg C by 2100. The countries most vulnerable to BC losses include Indonesia, Nigeria, Cuba, United States, Malaysia, Mexico, Australia, Philippines, Cameroon, and Papua New Guinea. Globally, BC gains and losses may be huge, on the order of −70 to 489 Tg C under RCP8.5 and −27 to 28 Tg C under RCP4.5 (Lovelock and Reef 2020).

While these model projections are very useful in estimating effects of (re-) establishment efforts and sustainable development scenarios, they do not encompass site variability which can be significant and must not be ignored. An analysis of CO_2 and CH_4 fluxes from a *Avicennia germinans* forest in the Tampamachoco lagoon, Mexico (Romero-Uribe et al. 2022), showed that there was a combined effect of seasonality and degradation on CH_4 fluxes, with highest fluxes during the rainy season in a dead mangrove stand. The main drivers of CH_4 fluxes were seasonal porewater salinity and NO_2^-, NO_3^- and SO_4^{-2} availability. CO_2 fluxes were highest in the dry season with no significant differences among degradation levels; major drivers of CO_2 fluxes were water level, redox potential, and sediment Cl^{-1} and $SO_4^{\ 2-}$ concentrations.

These model and empirical findings suggest that management of mangrove forests, aiming at their sustainable development as ecosystem service providers (BC, blue economy), must be both adaptive and region-, site- and community-specific.

7.3.2 CO_2 Emissions: Multiple Drivers

Deforestation and sea-level rise have not been the major drivers of mangrove loss since 2000. The major driver has been agriculture and aquaculture, having caused

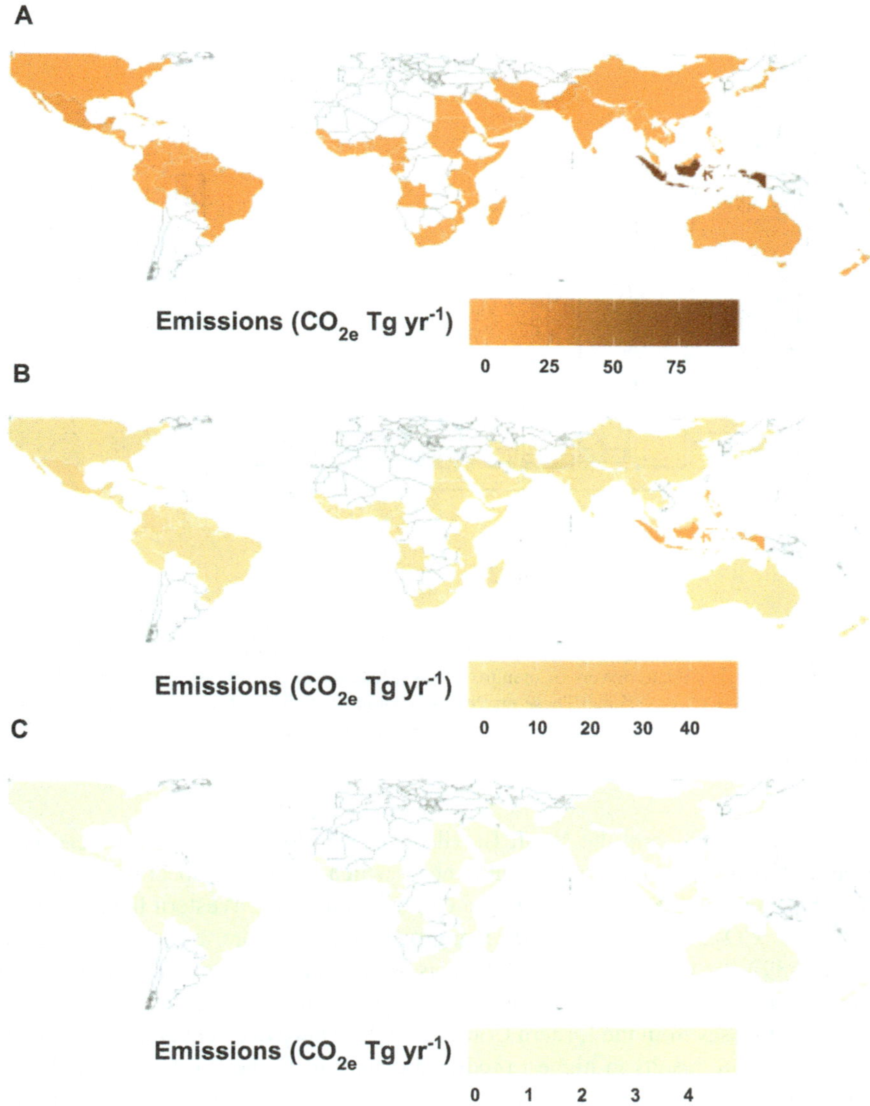

Fig. 7.2 Decline in global CO_{2eq} emissions from mangrove deforestation from (**A**) 1980, (**B**) 1990, and (**C**) 2000. Source: Chatting et al. (2022). Reproduced under a Creative Commons attribution 4.0 international license. CO_{2eq} for C stocks were estimated using a CO_2:C stoichiometric ratio of 3.67 (i.e., $CO_2/C = 44/12$)

conversion of 219,392 ha of mangroves from 2000–2016 (Adame et al. 2021), followed by erosion, which caused loss of 92,787 ha, mostly on the North Brazil Shelf, the Bay of Bengal, and the Sunda Shelf. The third driver has been extreme weather events, causing loss of 41,525 ha, mainly on the Sahul Shelf, tropical

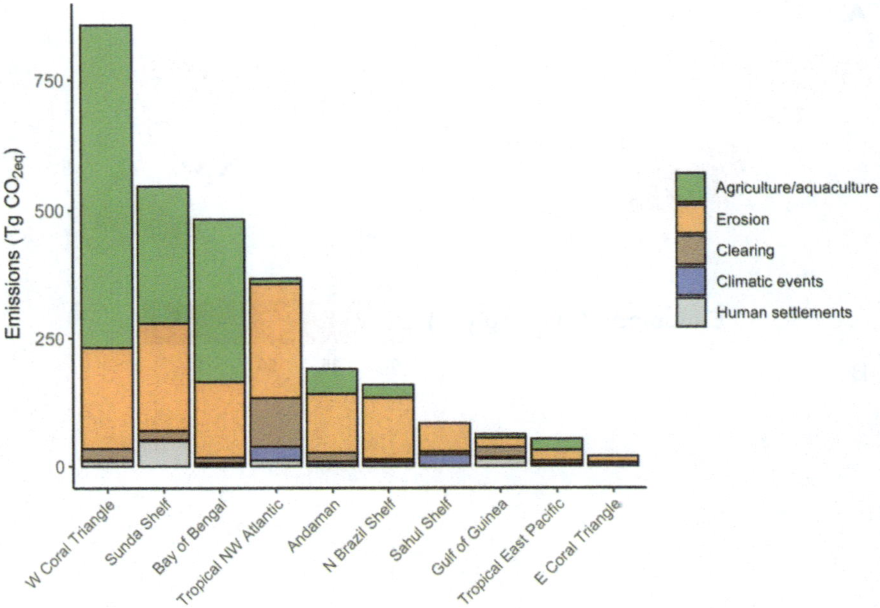

Fig. 7.3 Cumulative CO_{2eq} emissions (Tg) by the end of the twenty-first century (2010–2100) attributed to the proximate drivers of mangrove forest loss for the marine provinces ranked in the top ten for future CO_2 emissions. These provinces account for $>$ 90% of the total projected global emissions. Source: Adame et al. 2021. Reproduced under a Creative Commons attribution 4.0 international license

Northwest Atlantic, and the North Brazil Shelf. The fourth and fifth drivers have been deforestation and human settlement which caused respective losses of 39,595 ha and 10,529 ha, mostly in the Gulf of Guinea and Western Indian Ocean.

Global CO_{2eq} emissions upon mangrove area loss are projected to reach 2,391–3,392 Tg CO_{2eq} by 2100. There is significant geographical variability in projected losses (Fig. 7.3), with highest emissions from the Western Coral Triangle and lowest losses from the Eastern Coral Triangle. Mangrove conversion to agriculture/aquaculture results in highest predicted emissions in the Western Coral Triangle, Sunda Shelf, and the Bay of Bengal at 985 Tg CO_{2eq}, contributing 73% to the total emissions predicted to follow area loss; erosion is also an important driver of mangrove loss in these provinces (23%, 38% and 30% of their emissions, respectively). The Andaman province (West Myanmar, Bangladesh, East India) lost 98 Tg CO_{2eq} (60% of its total emissions) due to erosion. The tropical northwest Atlantic was also an emissions hotspot with large emissions from erosion (191 Tg CO_{2eq}), deforestation (80 Tg CO_{2eq}), and extreme weather events (23Tg CO_{2eq}). Potential CO_{2eq} emissions from lost mangrove stands vary considerably among locations, including within a country (Fig. 7.4). In Brazil, for instance, CO_{2eq} emissions vary by different drivers in different states, with erosion being the largest source of emissions, followed by agriculture/aquaculture; conversion to human settlements

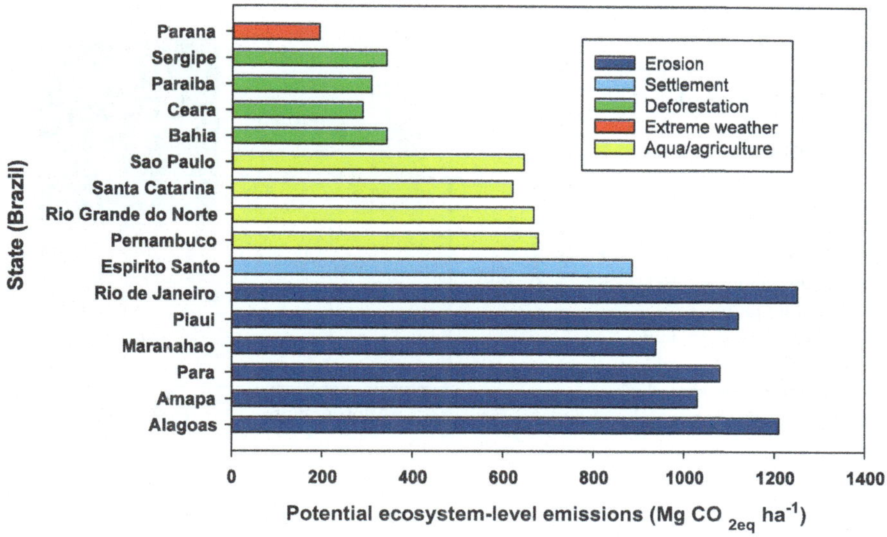

Fig 7.4 Potential total ecosystem CO_{2e} emissions in each state of Brazil grouped by driver of mangrove forest loss. Data from Table 5 in Rovai et al. (2022)

results in high emission rates but only in one state (Rovai et al. 2022). Surprisingly, extreme weather events and deforestation are projected to result in only low CO_{2eq} emissions.

7.3.3 Restoration and Conservation of Blue Carbon

A major component to sustainably manage mangrove forests is to conserve intact stands and to restore and rehabilitate degraded and converted forests. Alternatively, or additionally, suitable coastal habitats that emerge from sea level-rise or other climate change effects, can be established as mangrove stands through afforestation (e.g., Zimmer et al. 2022). If successful, ambitious global targets for (re-) establishment could deliver increased carbon sequestration, biodiversity, restoration of other ecosystem services, fisheries production, and protection of coastlines. However, most restoration projects fail for many reasons, including planting in unsuitable locations with inadequate community support, funding, and changing economic policies and plans (Lovelock et al. 2022; Zimmer et al. 2022). Capacity building and development is now seen as a vital part of successful (re-)establishment and conservation. Solutions to achieving success include minimizing risks of large projects and increasing the uptake and effectiveness of small projects.

Currently, there are over 370 mangrove restoration sites globally which show that reforestation (re-establishing sites previously inhabited by mangroves) has a greater

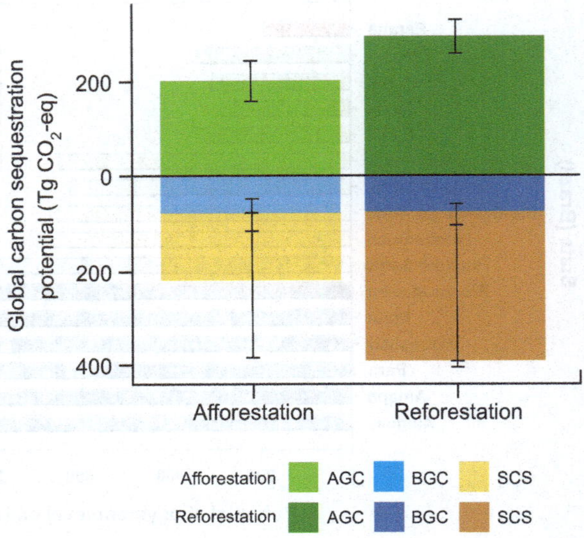

Fig. 7.5 The global maximum carbon mitigation potential by restoration and afforestation, modelled over a 40-year period (2021–2060) and assuming the same area for both reforestation scenarios. Error bars represent 95% confidence interval in each pool. Abbreviations: *AGC* aboveground biomass carbon, *BGC* belowground biomass carbon, *SCS* soil carbon. Source: Song et al. (2023)

C storage potential per hectare than afforestation (establishing mangroves where they did not previously exist). This result was attributed mostly to more favourable intertidal positioning, higher nitrogen availability, and lower salinity at most of the restoration sites compared to the afforested stands (Song et al. 2023). Future changes of coastlines and environmental conditions upon sea level-rise and warming might change this picture through changed habitat suitability. In a modelling scenario of carbon mitigation potential over 40 years (2021–2060) based on empirical data, greater C sequestration in reforested areas is vested in aboveground biomass C and soil C (Fig. 7.5), with no significant differences in belowground biomass C. Assuming all deforested mangrove areas are reforested within 1 year, the cumulative C sequestration over the next 40 years is 688.8 (624.8–752.7) Tg CO_{2eq}, which is roughly equivalent to other estimates in Sect. 7.3.1 and about 259 Tg CO_{2eq} higher than mangrove afforestation on tidal flats. Such reforestation would increase annual CO_2 uptake by mangrove ecosystems by 4.3–5.1% (Song et al. 2023).

Deforestation rates are declining, but conservation of remaining intact forests and/or forests of high intrinsic value are still urgently needed. Conserving existing mangroves is the most cost-effective and efficient measure for management of C sequestration. Some gains in mangrove areas have been detected in South America with a higher rate of gain through reforestation than loss , followed by North and Central America and the Caribbean and Southwest Asia (Worthington et al. 2020); about 39% of remaining mangroves reside within protected areas (PAs), especially in Brazil, Indonesia, and Mexico. However, the enforcement of PAs is very weak, ranging from limited protection to none. This is due to several factors, including poor governance by the state, conflict among management agencies, jurisdictional and regulatory ambiguities, and economic pressures (Ferreira et al. 2022).

Best conservation results are achieved in community-based protected areas, considering the increasing livelihood-sustaining relevance of mangrove resources, such as in Vietnam, the Philippines, and Ecuador where conservation is society-based rather than state-based.

Many indigenous communities have traditional practices of mangrove management and conservation integrated into their societal structures. Where communities are empowered and granted legal rights and authority to manage their own forests, the resulting community- based management has been effective in rationalizing the use of mangrove goods and services, as in some parts of Africa and Asia.

7.4 The Blue Economy in Sustainable Mangrove Development

The application of BC data for BC projects and the practical running of such a project for climate change-mitigation and mangrove re-establishment has progressed but not without its problems as reflected in low success rates of most projects (Alongi 2018; Zimmer et al. 2022). These problems have not stopped the push into what is called the 'blue economy', derived from Goal 14 of the UN Sustainable Development Goals to conserve and sustain marine resources. This goal reflects an upgrade on the traditional marine economy (e.g., oil and gas industry, mining, commercial fishing, aquaculture, shipping) to sustainable, clean technologies including maximizing BC. The new blue economy is to promote economic benefits of 'good for the ocean' industries and activities while ensuring truly sustainable development.

7.4.1 Financial Sustainability of Mangrove Blue Carbon

Much remains to be learned about BC offsetting, accrediting, and financing, and the development of finance schemes to fund projects remains a challenge. Several financial mechanisms such as the GEF Trust Fund, the Special Climate Change Fund, or the Green Climate Fund exist (Table 7.2), and BC projects can also be funded via the voluntary C market (VCM). The market involves risks and includes problems such as fluctuations in C prices.

The VCM deals with the selling and buying of emission reduction credits (offsets) in non- government-regulated markets. Mangrove C offset projects may be economically feasible at low to moderate prices of \$2 to \$11 US per t CO_{2eq} (Siikamäki et al. 2012), but the compliance market prices can be up to ten times higher than voluntary prices (Jacovac et al. 2020).

A recent analysis (Jacovac et al. 2020) of the C prices needed to promote mangrove conservation and re-establishment found that Southeast Asia holds the largest opportunities for BC projects, as conserving remaining mangroves would avoid the release of up to 15.5 Pg CO_2 to the atmosphere, with 90% of conservation

Table 7.2 Features of different funding approaches to blue carbon activities

Activity	Can occur in a developing (D) or developed (DV) country	Finance	C benefit
NAMAs/NAPAs[a]	D	DO, I	R
Climate-related ODA[b]	D	I	R
Bi- and multi-lateral activities[c]	D, DV	DO, I	R
REDD +	D	I	F
National NRM actions[d]	D, DV	DO	R
Voluntary offsets (e.g., VCS)[e]	D, DV	P	R, F
Compliance offsets (e.g., CDM)[f]	D	P	F
Domestic compliance offsets (e.g., CFI, CCERs)[g]	D, DV	P	R
CSR projects[h]	D, DV	P	R
Others (insurance microfinance, green bonds)[i]	D, DV	DO, I, P	R

Source: Alongi (2018). Reproduced with permission from Springer. Abbreviations: *DO* domestic public finance, *I* international public finance, *P* private, *R* remains, *F* flows

[a]National Appropriate Mitigation Actions (NAMAs) are agreed actions taken by developing countries as part of their commitments under the UNFCCC. National Adaptation Programmes of Action (NAPAs) are limited to least developed nations.

[b]Official development assistance (www.oecd.org/dac/)

[c]Bi- and multilateral activities (agreements between nations or regions or activities implemented through partnerships with public funding institutions such as World Bank or Asian Development Bank).

[d]National resource management can occur in many ways depending on local regulatory and social conditions.

[e]Voluntary market carbon offsets can be sourced through various providers including Verified C Standard, the American C Registry, etc.

[f]Regulated domestic emissions trading schemes require international carbon offsets sourced from mainly the Clean Development Mechanism (CDM) and joint implementation schemes established by the Kyoto Protocol.

[g]National C reduction schemes continue to be established. These legislative initiatives usually create their own domestic carbon offset units, generally oriented towards eventual integration with international market mechanisms.

[h]Corporate social responsibility (CSR), important area of BC funding often not considered in climate finance opportunities as many organisations may choose to invest in voluntary projects without a C focus.

[i]Climate bonds are a viable financial asset that can be issued by governments or privately, operating like standard debt instruments. There are three sources: national government, development of pilot and payment from verified emissions reductions.

being achieved at C prices between \$3 – \$13 US per t CO_2. It is the span between project costs (estimated \$0.5–2 per t CO_2) and gain from VCM that drives the attractiveness and feasibility of BC projects for investors. However, funding schemes of VCM, particularly for conservation projects, have recently been criticized for intransparency and leaving scope for green-washing or betrayal.

Improvement of project viability may likely include the need for other financial mechanisms and active participation of other sectors of society, as well as improved measures of monitoring, reporting, and verification (MRV).

Another pathway of funding is via payment for ecosystem services (PES). PES schemes offer scientific and policy challenges, but accurate valuations of ecosystem services offer incentives for funding and private investment as well as improved management and governance of these resources. Valuations often remain difficult to solve except on a case- by-case basis, as there are many ecosystem services that are interlinked to C, thus making it difficult to value one service over another. A number of problems exist concerning the 'blue economy'; the commoditisation of nature might encourage perverse outcomes, representing a shift from conservation motives to economic self-interest (Barbesgaard 2018).

7.4.2 Problems, Constraints and Limits

Successful implementation of BC projects is constrained by several economic (47%), political (24%), social (21%), and scientific (8%) issues; economic challenges were further classified as issues concerning funding (27%), people skills (22%), carbon price (14%), 'bad actors' (14%), high costs (14%) and other minor issues (11%) (Beeston et al. 2020).

Economic challenges include problems of sufficient seed funding, mismatched expectations, low or greatly fluctuating C price, lack of long-term funding, overly complex and expensive accreditation processes, human resources capacity, and inability to secure buyers for C offsets post-accreditation. Political issues involve land tenure rights and a mangrove policy gap where they are subject to overlapping or conflicting legislation. Social problems involve trust between local people and mangrove project developers, lack of effective stakeholder consultation, and complexity caused by multiple users, local social groups, and investors (Beeston et al. 2020).

A SWOT analysis of sustainable mangrove management in Indonesia (Table 7.3) offers a list of issues common to most BC projects globally. Five strategies formulated based on the opportunities and weaknesses identified in the SWOT analysis were: (1) greater cooperation between agencies and related parties; (2) create increased capacity-development of local governments to manage mangrove ecosystems in accordance with local conditions; (3) development of advanced study, science, technology, and information systems needed to enforce sustainability of mangrove ecosystems; (4) management of mangrove ecosystems via partnerships between government, local communities, and businesses with international institutional support to meet global environmental commitments; and (5) awareness-raising and community training to develop commodity processing from mangroves (Arifanti et al. 2022).

As mangroves are non-equilibrium ecosystems inhabiting a highly dynamic land-sea boundary, rarely conforming to the classic Clementsian concept of forest

Table 7.3 SWOT analysis of challenges and problems for sustainable mangrove management in Indonesia

Strengths	**Weaknesses**
Huge mangrove area (3.3 million hectares)	Policy, regulation, and implementation of laws still ego-sectoral
Mangroves have high biodiversity	Forest utilization still not in accordance with ecosystem carrying capacity
Mangroves have multiple ecological, economic, and social functions and benefits	Population size and pressure increasing
Regulations and laws exist related to mangrove forest management	Suboptimal support from institutions at the site level
Positive understanding of the community regarding conservation efforts	Weak monitoring, control, and evaluation by the government
They are one of Indonesia's natural resource assets	Low success rate of rehabilitation and restoration
Opportunities	**Threats**
Utilization supported by policies and regulations	Mangrove forest degradation
Benefits of high economic value	Exploitation of forest not according to land capability
Diversification of non-timber forest products (NTFP)	Decreased diversity of flora and fauna
National rehabilitation program	Loss or reduction of mangrove habitat
A harvest shelter that supports marketing of produce	Utilization of NTFPs without considering their sustainability
Access to transportation that supports produce marketing	Changes in land cultivation system
Rehabilitation technology and utilization pattern techniques	Climate change

Source: Arifanti et al. (2022). Reproduced under a Creative Commons attribution 4.0 international license

development and succession (Alongi 2011), several ecological issues thus present constraints for payment for ecosystem services (PES) and reducing emissions from deforestation and forest degradation (REDD+) schemes. For instance, while some VCM schemes account for the avoided loss of CO_2 upon conservation and prevented degradation, C payments are dependent on the rate of C sequestration and storage, not the size of C_{ORG} stocks. Thus, site selection should be confined to areas of highest sequestration which tends to be closest to the sea edge. Gas emissions and net ecosystem production are linked to forest age, especially in monospecific plantations. Further, while BC sequestration and storage is not necessarily higher in diverse than in low diversity stands (Rahman et al. 2021; Quadros et al. 2021), planting of mixed species forests is recommended when maximizing biodiversity is the goal of (re-) establishment, and BC schemes must give priority to maintain the existence of old-growth forests feeding into VCM budgets although BC-budgeting of avoided losses remains guess work and has been criticized repeatedly. Large uncertainties exist in both C sequestration potential and GHG emission rates of mangrove forests, and such limitations must be factored into the design, time frame, and execution of BC projects.

Even though mangrove forests are among the most efficient ecosystems in sequestering CO_2 and storing BC per unit area, their overall small global area renders their impact on climate change mitigation limited at the global scale. Taillardat et al. (2018) estimated that existing mangrove forests buffer only 0.42% of global fossil fuel emissions. Alongi (2020b) came to a similar conclusion that mangrove CO_2 emissions account for only 0.2% of total global emissions but estimated a contribution of 18% of emissions from the tropical coastal ocean. Reforesting the estimated 800,000 ha of previous mangrove area available and suitable for re-establishing mangroves (Worthington et al. 2020), and expanding beyond previous mangrove areas through afforestation, would significantly increase these estimated contributions. Both studies (Taillardat et al. 2018; Alongi 2020b) agree, however, that although mangrove forests play a limited role in climate change mitigation globally, their role is more significant within the tropics and at the national and regional scale, especially in areas of high rates of loss. Further, their societal value is exemplified by the multiple co- benefits they provide, rendering them attractive as multi-faceted nature-based solutions for climate change mitigation, ecosystem-based adaptation, and beyond.

There are also limits of C financing for mangrove forests. Only about 20% of mangrove forests can quality for BC financing, based on the probability of imminent threat of decline or loss if left unprotected by a conservation intervention (Zeng et al. 2021). Of these investible areas, 1.1–1.3 Mha can be financially sustainable over 30 years based on C prices of \$5–\$9.4 per t CO_{2eq}, contributing about 26 Mt CO_{2eq} a^{-1}, yielding a return on investment of about \$3.7 billion USD a^{-1}. Zeng et al. (2021) suggest that a disproportionately large potential of BC finance can be leveraged for national climate mitigation goals, especially in Brazil, Madagascar, Nigeria, Indonesia, Myanmar, Vietnam, and Papua New Guinea. Their analysis further indicates that information on return on investment underscores the potential for underutilized coastal BC projects in the tropics.

A transdisciplinary team of experts has recently concluded that challenges to BC management and marketing can be addressed by: improving policy and legal arrangements to ensure equitable benefit sharing; improving stewardship by incorporating knowledge and values of local communities; clarifying property rights; improving financial approaches and accounting tools to incorporate co-benefits; developing low cost, technological methods to measure and monitor carbon sequestration; and resolving knowledge gaps in BC biogeochemistry (Macreadie et al. 2022).

7.4.3 Lessons Learnt

The success of BC projects involves finding solutions to environmental, technical, social and societal, economic, and political issues and barriers. An analysis of four

BC projects in Kenya, India, Vietnam, and Madagascar (Wylie et al. 2016) highlights strengths and challenges of such mangrove projects that are illustrative of common issues. These projects demonstrated benefits of incorporating livelihood aspects into the design, and the importance of community involvement in planning and implementation along with awareness, ownership, stewardship, and other opportunities for income help to overcome threats due to local use of mangroves and their resources and to ensure that protecting a mangrove forest in one place does not lead to deforestation elsewhere. The most feasible funding mechanism for smaller projects seems to be the voluntary C market. Participants have a choice in methodologies for verifying C credits. The Verified Carbon Standard (VCS) and American Climate Registry (ACR) are often used globally to verify and issue C credits from field projects, to be traded on the voluntary C market (Alongi 2018). Projects can avoid the high costs and administrative burden associated with more formal funding mechanisms that adhere to the United Nations Framework Convention on Climate Change (UNFCCC). However, one of the downsides of the voluntary market is that credits are worth less than in the compliance market due to lower demand, differing quality standards, and lack of transferability (Wylie et al. 2016).

Additional lessons learnt include the need to incorporate sediment C into their C accounting, as well as measures for biodiversity conservation and local livelihood and wellbeing, and on the other hand, potential risks and offsets of BC projects. In the Sundarbans BC project, for instance, sediment C is monitored, but the possible effects of sea-level rise and delta subsidence are not. It is therefore possible that newly planted mangroves will drown if these changes are not accounted for over time. In the other three projects in Kenya, Vietnam and Madagascar, sediment C is not considered, missing most of the C to be accounted for to maximize their climate change-mitigation value. Leveraging resources elsewhere to better take future environmental changes into account would help greatly in maximizing the long-term success of BC projects. In Indonesia, a robust assessment found that several key factors often neglected to improve restoration success should be incorporated in BC projects, such as the impact of changes in geomorphology due to 20 years of land-use and land-cover change, and the status of state forest land (Sasmito et al. 2023). Further, improving monitoring and evaluation and government representation at the subnational level will increase the likelihood of reducing failure and achieving rehabilitation targets.

References

Adame MF, Connolly RM, Turshwell MP, Lovelock CE, Fatoyinbo T, Lagomasino D et al (2021) Future carbon emissions from global mangrove forest loss. Glob Change Biol 27:2856–2866. https://doi.org/10.1111/gbc.15571

Alongi DM (2009) The energetics of mangrove forests. Springer, Berlin

Alongi DM (2011) Carbon payments for mangrove conservation: ecosystem constraints and uncertainties of sequestration potential. Environ Sci Policy 14:462–470. https://doi.org/10.1016/j.envsci.2011.02.004

Alongi DM (2014) Carbon cycling and storage in mangrove forests. Annu Rev Mar Sci 6:195–219. https://doi.org/10.1146/annurev-marine-010213-135020
Alongi DM (2018) Blue carbon: coastal sequestration for climate change mitigation Springer briefs in climate studies. Springer Nature, Cham
Alongi DM (2020a) Carbon cycling in the world's mangrove ecosystems revisited: significance of non-steady state diagenesis and subsurface linkages between the forest floor and the coastal ocean. Forests 11:977. https://doi.org/10.3390/f11090977
Alongi DM (2020b) Global significance of mangrove blue carbon in climate change mitigation. Sci 2:67. https://doi.org/10.3390/sci2030067
Alongi DM (2022) Impacts of climate change on blue carbon stocks and fluxes in mangrove forests. Forests 13:149. https://doi.org/10.3390/f13020149.9
Arifanti VB, Sidik F, Mulyanto B, Susilowati A, Wahyuni T, Subarno et al (2022) Challenges and strategies for sustainable mangrove management in Indonesia: a review. Forests 13:695. https://doi.org/10.3390/f13050695
Barbesgaard M (2018) Blue growth: saviour or ocean grabbing? J Peasant Stud 45:130–149. https://doi.org/10.1080/03066150.2017.1377186
Beeston M, Cuyvers L, Vermilye J (2020) Blue carbon: mind the gap. Gallifrey Foundation, Geneva
Belliard JP, Hernandez S, Temmerman S, Suello RH, Dominguez-Granda LE, Rosado- Moncayo AM et al (2022) Carbon dynamics and carbon dioxide and methane exchange in the mangrove dominated Guayas River delta, Ecuador. Estuar Coast Shelf Sci 267:107766. https://doi.org/10.1016/j.ecss.2022.107766
Bertram C, Quaas M, Reusch TB, Vafeidis AT, Wolff C, Rickels W (2021) The blue carbon wealth of nations. Nat Clim Change 11:704–709. https://doi.org/10.1038/s41588-021-01089-4
Carnell PE, Palacios MM, Waryszak P, Trevathan-Tackett SM, Masqué P, Macreadie PI (2022) Blue carbon drawdown by restored mangrove forests improves with age. J Environ Manag 306:114301. https://doi.org/j.envman.2021.114301
Chatting M, Al-Maslamani I, Walton M, Skov MW, Kennedy H, Husrevoglu YS et al (2022) Future mangrove carbon storage under climate change and deforestation. Front Mar Sci 9:781876. https://doi.org/10.3389/fmars.2022.781876
Chen G, Gao M, Pang B, Chen S, Ye Y (2018) Top-meter sediment organic carbon stocks and sources in restored mangrove forests of different ages. For Ecol Manag 422:87–94. https://doi.org/10.1016/j.foreco.2018.03.044
Cheriyan AS, Moushmi KS, Cheriyan E, Baby L, Chandramohanakumar N (2022) Sources and degradation of organic matter and its relation to trophic status in the core sediments of a tropical mangrove ecosystem along the southwest coast of India. Mar Chem 244:104137. https://doi.org/10.1016/j.marchem.2022.104137
Cohen MCL, Pessenda LCR, Behling H, de Fátima Rossetti D, França MC, Guimarães JTF et al (2012) Holocene paleoenvironmental history of the Amazonian mangrove belt. Quat Sci Rev 55:50–58. https://doi.org/10.1016/j.quascirev.2012.08.019
Collins DS, Nguyen VL, Ta TKO, Mao L, Ishii Y, Ktagawa H et al (2021) Sedimentary evolution of a delta-margin mangrove in Can Gio, northeastern Mekong River delta, Vietnam. Mar Geol 433:106417. https://doi.org/10.1016/j.margeo.2020.106417
Cormier N, Krauss KW, Demopoulos AWJ, Jessen BJ, McClain-Counts JP, From AS et al (2022) Potential for carbon and nitrogen sequestration by restoring tidal connectivity and enhancing sediment surface elevations in denuded and degraded South Florida mangrove ecosystems. In: Krauss KW, Zhu Z, Stagg CL (eds) Wetland carbon and environmental management, American Geophysical Union. Wiley, Hoboken, pp 143–158. https://doi.org/10.1002/978111.963935.ch7
Dontis EE, Radabaugh KR, Chappel AR, Russo CE, Moyer RP (2020) Carbon storage increases with site age as created salt marshes transition to mangrove forests in Tampa Bay, Florida (USA). Estuaries Coast 43:1470–1488. https://doi.org/10.1007/s122237-020-00733-0

Elmstrom E, Valiela I, Fox SE (2018) Watershed-estuary coupling in Pacific Panama: isotopic evidence of forest and pastureland covers on watersheds and marine contributions to suspended particulate matter in mangrove estuaries. Estuaries Coast 41:1968–1976. https://doi.org/10.1007/s12237-018-0400-0

Ferreira AC, Borges R, de Lacerda LD (2022) Can sustainable development save mangroves? Sustainability 14:1263. https://doi.org/10.3390/su14031263

Friess D, Yando ES, Abuchahla GMO, Adams JB, Cannicci S, Canty SWJ et al (2020) Mangroves give cause for conservation optimism, for now. Curr Biol 30:R135–R158

Goldberg L, Lagomasino D, Thomas N, Fatoyinbo T (2020) Global declines in human- driven mangrove loss. Glob Change Biol 26:5844–5855. https://doi.org/10.1111/gcb.1575

Hamzeh MA, Lahijani HAK (2022) Sediment and vegetative carbon sequestration in Khuran estuary mangroves, Strait of Hormoz, during the last 18 centuries. Estuaries Coast 45:1583–1595. https://doi.org/10.1007/s12237-021-01037-7

Hanggara BB, Murdiyarso D, Ginting YRS, Widha YL, Panjaitan GY, Lubis AA (2021) Effects of diverse mangrove management practices on forest structure, carbon dynamics and sedimentation in North Sumatra, Indonesia. Estuar Coast Shelf Sci 259:107467. https://doi.org/10.1016/j.ecss.2021.107467

Hargan KE, Williams B, Nuangsaeng B, Siriwong S, Tassawad P, Chaiharn C et al (2020) Understanding the fate of shrimp aquaculture effluent in a mangrove ecosystem: aiding management for coastal conservation. J Appl Ecol 57:754–765. https://doi.org/10.1111/1365-2664.13579

Hatje V, Masqué P, Patire VF, Dórea BF (2021) Blue carbon stocks, accumulation rates, and associated spatial variability in Brazilian mangroves. Limnol. Oceanogr. 66:321–324. https://doi.org/10.1002/lno.11607

Hilmi N, Chami R, Sutherland MD, Hall-Spencer JM, Lebleu L, Benitez MB et al (2021) The role of blue carbon in climate change mitigation and carbon stock conservation. Front Clim 3: 710546. https://doi.org/10.3389/fclim.2021.710546

Jacovac CC, Latawiec AE, Lacerda E, Lucas IL, Korys KA, Iribarrem A et al (2020) Costs and carbon benefits of mangrove conservation and restoration: a global analysis. Ecol Econ 176: 106758. https://doi.org/10.1016/j.ecolecon.2020.106758

Jia M, Wang Z, Mao D, Ren C, Song K, Zhao C et al (2023) Mapping global distribution of mangrove forests at 10-m resolution. Sci Bull. https://doi.org/10.1016/j.scib.2023.05.004

Kandasamy K, Rajendran N, Balakrishnan B, Thiruganasambandam R, Narayanasamy R (2021) Carbon sequestration and storage in planted mangrove stands of *Avicennia marina*. Reg Stud Mar Sci 43:101701. https://doi.org/10.1016/j.rsma.2021.101701

Kathiresan K, Anburaj R, Gomathi V, Saravanakumar K (2013) Carbon sequestration potential of *Rhizophora mucronata* and *Avicennia marina* as influenced by age, season, growth and sediment characteristics in southeast coast of India. J Coast Conserv 17:397–408. https://doi.org/10.1007/s11852-013-0236-5

Kristensen E, Bouillon S, Dittmar T, Marchand C (2008) Organic carbon dynamics in mangrove ecosystems: a review. Aquat Bot 89:201–219. https://doi.org/10.1016/j.aquabot.2007.12.005

Kristensen E, Connolly RM, Otero XL, Marchand C, Ferreira TO Rivera-Monroy VH (2017) Biogeochemical cycles: global approaches and perspectives. In: Rivera-Monroy VH, Lee SY, Kristensen E, Twilley RR (eds) Mangrove ecosystems: a global and biogeographic perspective. Springer, Cham, Switzerland, pp 163–209. https://doi.org/10.1007/978-3-319-62206-4_1

Kusumaningtyas MA, Hutahean AA, Fischer HW, Pérez-Mayo M, Ransby D, Jennerjahn TC (2019) Variability in the organic carbon stocks, sources, and accumulation rates of Indonesian mangrove ecosystems. Estuar Coast Shelf Sci 218:310–323. https://doi.org/10.1016/j.ecss.2018.12.007

Kusumaningtyas MA, Kepel TL, Solihuddin T, Lubis AA, Putra ADP, Sugiharto U et al (2022) Carbon sequestration potential in the rehabilitated mangroves in Indonesia. Ecol Res 37:80–91. https://doi.org/10.1111/1440-1703.12279

Le Minor M, Bartzke G, Zimmer M, Gillis L, Helfer V, Huhn K (2019) Numerical modelling of hydraulics and sediment dynamics around mangrove seedlings: implications for mangrove establishment and reforestation. Estuar Coast Shelf Sci 217:81–95. https://doi.org/10.1016/j.ecss.2018.10.019

Liu T, Liu S, Wu B, Xu H, Zhang H (2020) Increase in organic carbon burial response to mangrove expansion in the Nanliu River estuary, South China Sea. Prog Earth Planet Sci 7:71. https://doi.org/10.1186/s40645-020-00387-3

Locatelli T, Binet T, Kairo JG, King L, Madden S, Patenaude G et al (2014) Turning the tide: how blue carbon and payments for ecosystem services (PES) might help save mangrove forests. Ambio 43:981–995. https://doi.org/10.1007/s13280-014-0530-y

Lovelock CE, Reef R (2020) Variable impacts of climate change on blue carbon. One Earth 3:195–211. https://doi.org/10.1016/j.oneear.2020.07.010

Lovelock CE, Ruess RW, Feller IC (2011) Carbon dioxide efflux from cleared mangrove peat. PloS One 6:e21279. https://doi.org/10.1371/journal.pone.0021279

Lovelock CE, Barbier E, Duarte CM (2022) Tackling the mangrove restoration challenge. PLoS Biol 20:e3001836. https://doi.org/10.1371/journal.pbio.3001836

MacKenzie R, Sharma S, Rovai AR (2021) Environmental drivers of blue carbon burial and sediment carbon stocks in mangrove forests. In: Sidik F, Friess DA (eds) Dynamic sedimentary environments of mangrove coasts. Elsevier, Amsterdam, pp 275–294. https://doi.org/10.1016/B978-0-12-816437-2.00006-9

Macreadie PI, Robertson AI, Spinks B, Adams MP, Atchison JM, Bell-James J et al (2022) Operationalizing marketable BC. One Earth 5:485–492. https://doi.org/10.1016/j.oneear.2022.04.005

Maher DT, Santos IR, Schulz KG, Call M, Jacobsen GE, Sanders CJ (2017) Blue carbon oxidation revealed by radiogenic and stable isotopes in a mangrove system. Geophys Res Lett 44:4889–4896. https://doi.org/10.1002/2017GL073753

Malhi Y, Doughty C, Galbraith D (2011) The allocation of ecosystem net primary productivity in tropical forests. Philos Trans R Soc B Biol Sci 366(1582):3225–3245. https://doi.org/10.1098/rstb.2011.0062

Matos CR, Berrêdo JF, Machado W, Sanders CJ, Metzger E, Cohen MC (2020) Carbon and nutrient accumulation in tropical mangrove creeks, Amazon region. Mar Geol 429:106317. https://doi.org/10.1016/j.margeo.2020.106317

Minu A, Routh J, Machiwa JF (2020) Distribution and sources of organic matter in the Rufiji Delta in Tanzania: variability and environmental implications. Appl Geochem 122:104733. https://doi.org/10.1016/j.apgeochem.2020.104733

Ouyang X, Kristensen E, Thornber C, Zimmer M, Yang Z, Lee SY (2023) Response of macrophyte litter decomposition in global blue carbon ecosystems to climate change. Global Change Biol 00:1–15. https://doi.org/10.1111/gcb.16693

Passos T, Penny D, Sanders C, De França E, Oliveira T, Santos L et al (2021) Mangrove carbon and nutrient accumulation shifts driven by rapid development in a tropical estuarine system, northeast Brazil. Mar Pollut Bull 166:112219. https://doi.org/10.1016/j.marpolbul.2021.112219

Perdomo Trujillo LV, Mancera-Pineda JE, Medina-Calderon JH, Zimmer M (2021) Massive loss of aboveground biomass and its effect on sediment organic carbon concentration: less mangrove more carbon? Estuar Coast Shelf Sci:248106888. https://doi.org/10.1016/j.ecss.2020.106888

Quadros AF, Nordhaus I, Reuter H, Zimmer M (2019) Modelling of mangrove annual leaf litterfall with emphasis on the role of vegetation structure. Estuar Coast Shelf Sci 218:292–299. https://doi.org/10.1016/j.ecss.2018.12.012

Quadros AF, Helfer V, Nordhaus I, Reuter H, Zimmer M (2021) Functional traits of terrestrial plants in the intertidal: A review on mangrove trees. Biol Bull 241:123–139. https://doi.org/10.1086/716510

Rahman MM, Zimmer M, Ahmed I, Donato D, Kanzaki M, Xu M (2021) Co-benefits of protecting mangroves for biodiversity conservation and carbon storage. Nat Commun 12:3875. https://doi.org/10.1038/s41467-021-24207-4

Rani V, Nandan SB, Schwing PT (2021) Carbon source characterization and historical carbon burial in three mangrove ecosystems on the southwest coast of India. Catena 197:104980. https://doi.org/10.1016/j.catena.2020.104980

Ratul SB, Gu X, Qiao P, Sagala FW, Nan S, Islam N et al (eds) (2022) Blue carbon sequestration following mangrove restoration: evidence from a carbon neutral case in China. Ecosyst Health Sust 8:2101547. https://doi.org/10.1080/20964129.2022.2101547

Ray R, Suwa R, Miyajima T, Munar J, Yoshikai M, San-Diego ML et al (2022) Sedimentary blue carbon dynamics based on chrono-sequential observation in a tropical restored mangrove forest. Biogeosci. https://doi.org/10.5194/bg-2021-359

Romero-Uribe HM, López-Portillo J, Reverchon F, Hernández ME (2022) Effect of degradation of a black mangrove forest on seasonal greenhouse gas emissions. Environ Sci Pollut Res 29: 11951–11965. https://doi.org/10.1007/s11356-021-16597-1

Rosentreter JA, Al-Haj AN, Fulweiler RW, Williamson P (2021) Methane and nitrous oxide emissions complicate blue carbon assessments. Glob Biogeochem. Cycles 35: e2020GB006858. https://doi.org/10.1029/2020GB006858

Rovai AS, Twilley RR, Worthington TA, Riui P (2022) Brazilian mangroves: blue carbon hotspots of national and global relevance to national climate solutions. Front For Glob Change 4:787533. https://doi.org/10.3389/ffgc.2021.787533

Salmo SG III, Gianan ELD (2019) Post-disturbance carbon stocks and rates of sequestration: implications on 'blue carbon' estimates of Philippine mangroves. Philipp Sci Lett 12:122–132

Sasmito SD, Kuzyakov Y, Lubis AA, Murdiyarso D, Hutley LB, Bachri S et al (2020) Organic carbon burial and sources in sediments of coastal mudflat and mangrove ecosystems. Catena 187:104414. https://doi.org/10.1016/j.catena.2019.104414

Sasmito SD, Basyuni M, Kridalaksana A, Saragi-Sasmito MF, Lovelock CE, Murdiyarso D (2023) Challenges and opportunities for achieving Sustainable Development Goals through restoration of Indonesia's mangroves. Nature Evol Evol 7:62–70. https://doi.org/10.1038/s41559-022-01926-5

Siikamäki J, Sanchirico JN, Jardine SL (2012) Global economic potential for reducing carbon dioxide emissions from mangrove loss. Proc Nat Acad Sci 109:14369–14374. https://doi.org/10.1073/pnas.120059109

Song S, Ding Y, Li W, Men Y, Zhou J, Gou R et al (2023) Mangrove reforestation provides greater blue carbon benefit than afforestation for mitigating global climate change. Nat Commun 14: 756. https://doi.org/10.1038/s41467-023-36477-1

Soper FM, MacKenzie RA, Sharma S, Cole TG, Litton CM, Sparks JP (2019) Non-native mangroves support carbon storage, sediment carbon burial, and accretion of coastal ecosystems. Glob Change Biol 25:4315–4326. https://doi.org/10.1111/gcb.14813

Suello RH, Hernandez SL, Bouillon S, Belliard J-P, Dominguez-Granda L, Van de Broek M et al (2022) Mangrove sediment organic carbon storage and sources in relation to forest age and position along a deltaic salinity gradient. Biogeosci 19:1571–1585. https://doi.org/10.5194/bg-19-1571-2022

Taillardat P, Friess DA, Lupascu M (2018) Mangrove blue carbon strategies for climate change mitigation are most effective at the national scale. Biol Lett 14:20180251. https://doi.org/10.1098/rsbl.2018.0251

Wigand C, Eagle M, Branoff BL, Balogh S, Miller KM, Martin RM et al (2021) Recent carbon storage and burial exceed historic rates in the San Juan Bay estuary peri-urban mangrove forests (Puerto Rico, United States). Front For Glob Change 4:676691. https://doi.org/10.3389/ffgc.2021.676691

Worthington TA, Andradi-Brown DA, Bhargava R, Buelow C, Bunting P, Duncan C et al (2020) Harnessing big data to support the conservation and rehabilitation of mangrove forests globally. One Earth 2:429–443. https://doi.org/10.1016/j.oneear.2020.04.0 8

Wylie L, Sutton-Grier AE, Moore A (2016) Keys to successful blue carbon projects: lessons learned from global case studies. Mar Policy 65:76–84. https://doi.org/10.1016/j.marpol.2015.12.020

Xiong Y, Liao B, Wang F (2018) Mangrove vegetation enhances sediment carbon storage primarily through in situ inputs rather than increasing allochthonous sediments. Mar Pollut Bull 131:378–385. https://doi.org/10.1016/j.marpolbul.2018.04.043

Zeng Y, Friess DA, Sarira TV, Siman K, Koh LP (2021) Global potential and limits of mangrove blue carbon for climate change mitigation. Curr Biol 31:1737–1743. https://doi.org/10.1016/j.cub.2021.01.070

Zhang Y, Meng X, Xia P, Zhang J, Liu D, Li X (2021) Spatiotemporal variations in the organic carbon accumulation rate in mangrove sediments from the Yingluo Bay, China, since 1900. Acta Oceanol Sinica 40:65–77. https://doi.org/10.1007/s13131-021-1864-5

Zhang Z, Wang Y, Zhu Y, He K, Li T, Mishra U et al (2022) Carbon sequestration in sediment and biomass under native and non-native mangrove ecosystems. Plant Soil. https://doi.org/10.1007/s11104-022-05352-1

Zimmer M, Helfer V (2020) Quantity and quality of organic matter in mangrove sediments. In: Sidik F, Friess D (eds) Dynamic sedimentary environments of mangrove coasts. Elsevier, Amsterdam, The Netherlands, pp 369–391. https://doi.org/10.1016/B978-0-12-816437-2.00015-x

Zimmer M, Ajonina GN, Amir AA, Cragg S, Crooks S, Dahdouh-Guebas F et al (2022) When nature needs a helping hand: different levels of human intervention for mangrove (re-) establishment. Front For Glob Change 5:784322. https://doi.org/10.3389/ffgc.2022.784322

Chapter 8
Community-Based Mangrove Restoration on Príncipe Island

António D. Abreu, A. Herrero Barrencua, and R. Haroun

Abstract A preliminary characterisation of Praia Salgada mangrove in Principe Island, Democratic Republic of São Tomé e Príncipe, was carried out to establish a baseline for intervention in restoring the mangrove habitat and support the development of sustainable tourism activities as an alternative income to the local community. The general characterisation of the mangrove of Praia Salgada included the identification of different types of vegetation cover, the recording of 24 species of birds, 13 species of molluscs, 11 species of crustaceans and 13 species of fish. The physicochemical parameters measured were pH, oxygen, salinity, turbidity and temperature.

The UNESCO Biosphere Reserve promoted the presentation and discussion of the results with the community, which resulted in the willingness to promote the restoration of the mangrove and the exploitation of the area to promote nature tourism activities, namely hiking trails and bird and plant watching. In this way, the community benefits from the availability of mangrove ecosystem services to improve their economic income and alternative jobs. It strengthens their commitment and involvement in the ecological restoration of this vital ecosystem.

Keywords Biosphere reserve · Community-based · Mangrove · UNESCO · Restoration sustainable tourism

A. D. Abreu (✉)
UNESCO Chair in Biodiversity Conservation for Sustainable Development, University of Coimbra, Coimbra, Portugal

A. H. Barrencua · R. Haroun
Biodiversity and Conservation Research Group, Institute ECOAQUA, University of Las Palmas de Gran Canaria, Las Palmas, Spain
e-mail: ricardo.haroun@ulpgc.es

M. Clüsener-Godt et al. (eds.), *Blue Carbon Mangrove Ecosystems*, Blue Carbon Ecosystems for Sustainable Development 1,
https://doi.org/10.1007/978-3-031-69553-7_8

8.1 Introduction

Principe Island is approximately 225 kilometres (140 miles) north of São Tomé Island and its terrestrial coverage is about 148 square kilometres. The island is of volcanic origin, characterised by lush rainforests, unique geological formations pristine beaches and large undisturbed coastline stretches.

Principe Island was nominated as UNESCO Biosphere Reserve in 2012, aiming to enhance its remarkable biodiversity.

Although at present time, mangrove is a type of habitat only found in 3 sites along the coastlines of Príncipe Island, they play a crucial role in the island's ecosystem. These salt-tolerant trees act as a buffer between the land and sea, protecting the coastline from erosion, storm surges, and tsunamis. Additionally, they serve as nurseries for various marine species, including fish and crustaceans, providing a safe environment for them to breed and grow. Mangroves also contribute to carbon sequestration, helping mitigate the impacts of climate change.

Due to their small size and disturbances suffered through human action, mangroves on Príncipe Island are in a critical state, making the need for recovery and conservation urgent. The UNESCO Biosphere Reserve has as one of its main objectives to promote biodiversity conservation and sustainable use. It has therefore enabled an initiative to restore the mangrove of Praia Salgada with the active participation of the community and exploring the possibility of associating nature tourism activities in the mangrove as a tool for job creation and diversification of the local micro-economy.

8.2 Methodology

The study of the Praia Salgada mangrove was carried out over a total area of 5152 hectares, starting at the mouth of the estuary that flows into the beach and going up the riverbed for almost 400 m, studying the first 50 metres of each bank to characterise the ecosystems that make up this biome. A sampling scheme was implemented to map and characterise this area's biodiversity and main physico-chemical parameters (Fig. 8.1).

The results were presented to the community and supported a participatory approach leading to the development of a community-based restoration of the mangrove associated with establishing a trail and birdwatching facilities to be managed by the community on nature-based tourism activities (Fig. 8.2).

Fig. 8.1 Principe Island geographic location and map

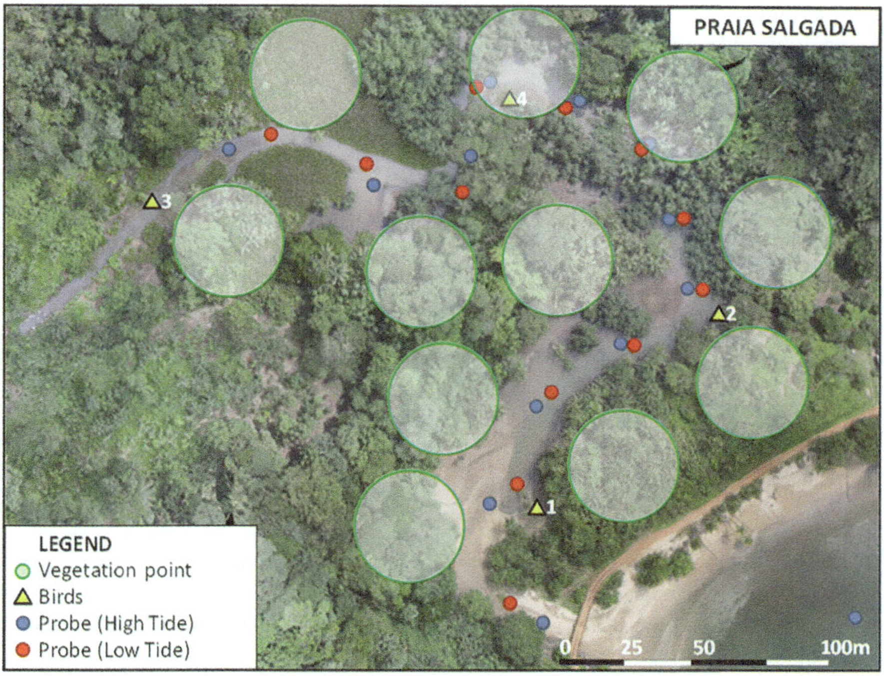

Fig. 8.2 Aerial view of Praia Salgada mangrove swamp and sampling stations. (Photo, courtesy of Felipe Spina)

8.3 Results

8.3.1 Vegetation

In Table 8.1 it is described the vegetation types identified inside the mangrove forest of Praia Salgada, including the area they occupy (hectares, Ha) and the corresponding percentage.

The tree cover is a luxuriant canopy, which gives rise to a leafy and very leafy canopy in 73.8% of the area studied, with tree species with a broad and lush canopy so that the ground under-cover tends to be low, with little colonisation by herbaceous strata (72.8%). In the remaining 23.2%, where tree cover is scarce, herbaceous species are abundant, predominantly *Acrostichum aureum* and water grasses.

The mangrove in this area is distributed in three zones, with the largest area covering almost the entire habitat, with mature individuals up to 20 m tall and a maximum trunk diameter of up to 12 cm that forms an outline of approximately 38 cm. Individuals constantly produce foliferous buds in terminal position and morphologically naked (Vargas 2011). The first flowering buds began to appear in September, although they do not have flowers before January. The reproductive

Table 8.1 Characteristics of the habitat's types identified in Praia Salgada, with the area occupied (Ha) and the percentage of the study area they represent

Vegetation	Description	Ha	%
Mangrove	Area inundated or flooded by salt water with mangrove (*Rhizophora harrisonii*) tree cover.	0,672	13,045
Herbaceous marsh	Flooded or floodable areas without tree species, predominantly herbaceous or shrub layer and possible areas of exposed sediment.	0,688	13,354
Flood coconut groves	Surrounded and flooded river areas with tree layers dominated by coconut palms (*Cocos nucifera*).	0,943	18,304
Flooded palm grove	Surrounded and flooded river areas with tree layers dominated by palms (*Elaeis guineensis*).	0,153	2970
Flood forest	Surrounded and flooded river areas where the tree layer is diversified, composed of different species and none of them predominant.	0,619	12,015
Coconut grove	Non-flooded area with tree layer dominated by coconut palms (*Cocos nucifera*).	0,221	4292
Palm grove	Non-flooded area with tree layer dominated by palm trees (*Elaeis guineensis*).	0,105	2038
Fields	Non-flooded area of small trees where horticulture is maintained.	0	0
Secondary forest	Non-flooded areas of fruit trees without agricultural use, as regeneration resulting from the removal of the original vegetation by human activity.	0,664	12,888
Bushes	Non-flooded area with a predominance of shrubs and low trees.	0,487	9452
Water	Area covered by water.	0,600	11,642
TOTAL		**5152**	**100**

fruit, elongated and approximately 20 cm long, begins to be seen discreetly at the end of November and the beginning of December. The native population reports that the buds flower around April and May, and the fruit ripens and falls around the middle of the year.

8.3.2 Birds

Six hundred eighty-eight bird specimens belonging to 24 different species were registered, giving an average of 43 birds per 10 minutes of sampling and 19.5 species per sighting point. Of all these sightings, 394 birds (57.27%) are endemic species, while 77 (11.2%) are endemic subspecies (Table 8.2).

Table 8.2 List of bird species observed in the Mangrove forest of Praia Salgada

Common name	Scientific name	Endemism	Sta.	Ab.	Freq.
Andorinhão de barriga branca	*Zoonavena thomensis*	Species (STP)	LC	67	100
Andorinhão de palmeira	*Cypsiurus parvus*	Species (STP)	LC	36	75
Céssia	*Treron sanctithomae*	No	VU	16	100
Chó-chó	*Halcyon malimbica dryas*	Species (STP)	LC	17	100
Conóbia	*Alcedo cristata thomensis*	Subspecies (P)	–	44	100
Currucucu	*Streptopelia senegalensis*	Subspecies (ST)	LC	11	50
Estorninho do Príncipe	*Lamprotornis ornatus*	No	LC	2	25
Estorninho bôbô	*Lamprotornis splendidus*	Species (STP)	LC	167	100
Falcão	*Milvus migrans*	No	LC	13	75
Garça boieira (braca)	*Bubulcus ibis*	No	LC	4	50
Garça boieira (preta)	*Bubulcus ibis?*	No	LC	5	50
Garça dos recifis	*Egretta gularis*	No	LC	2	50
Merlo	*Ploceus princeps*	No	LC	89	100
Munquê	*Columba larvata principalis*	Species (STP)	LC	10	75
Papagaio	*Psittacus erithacus*	Subspecies (ST)	EN	22	100
Papatela large	*Numenius phaeopus*	No	LC	18	100
Papatela pequeno	*Actitis hypoleucos*	No	LC	25	100
Peito branco	*Speirops leucophaeus*	No	NT	42	100
Rola	*Columba malherbii*	Species (P)	NT	12	100
Tchibi barbeiro	*Anabathmis hartlaubii*	Species (STPA)	LC	22	100
Tchibi peito branco	*Horizorhinus dohrni*	Species (STP)	LC	32	100
Tchibi singa	*Lonchura cucullata*	Species (STP)	LC	24	100
Tchonzu	*Butorides striata*	Species (STP)	LC	7	75
Sp. no ID				1	25
TOTAL				**688**	

Endemism indicates whether the species is endemic to the islands of the Gulf of Guinea: São Tomé (ST), Principe (P), Anobom (A), and species or subspecies level (Melo and Jones 2008). Sta. refers to the conservation category according to the International Union for Conservation of Nature Red List (IUCN 2017): NE: Not Evaluated, LC: Least Concern, NT: Near Threatened, VU: VUlnerable or EN: ENdangered. Ab. indicates the number of specimens recorded. Freq. indicates the percentage of sampling points where the species has been sighted.

8.3.3 Benthic macrofauna

Through direct observation and the collection of specimens for subsequent study in the laboratory, at least 24 species of animals classified as macrofauna have been observed in the Praia Salgada mangrove swamp, divided into 13 species belonging to the phylum Mollusca and another 11 species belonging to crustaceans (subphylum of arthropods).

The 13 mollusc species belong to 10 families, of which 7 are snails and 6 are bivalves. Within the bivalve group, 3 are oyster species, and 3 are different species of clam, which, together with the buzio de terra (a highly prized and coveted resource on the island), constitute an essential part of the daily diet in coastal areas.

Concerning the 11 species of crustaceans belonging to 8 different families, except for one species of shrimp, all the others belong to different families of crabs, several of them being an important food resource for the population (Table 8.3).

In the case of crack, which is considered a delicacy by the regional population, it is a typical dish of the local gastronomic culture. It is frequently eaten at festivals and popular celebrations. The various bivalve species form part of the daily diet along with the fisheries of coastal communities. Locals speak of other species(s) of shrimp being harvested for consumption, but this was not recorded during the present study.

8.3.4 Fishes

13 fish species belonging to 8 families have been recorded in the Praia Salgada mangrove swamp. Of the 13 species, 8 are commercial for human consumption, and the sole species *Arnoglossus imperialis, is the* first observation and citation for São Tomé and Príncipe (Table 8.4).

The term phase corresponds to the age of the individuals, with J (juvenile) being the state of sexual immaturity and A (adult) individuals of reproductive age according to their gonads.

Table 8.3 List of benthic macrofauna organisms registered at Praia Salgada

Taxa	Family	Species	Common name
Crustacea	Coenobitidae	*Coenobita rubescens*	Eremita de terra
Crustacea	Gecarcinidae	*Cardisoma armatum*	Carangejo de mangue
Crustacea	Gecarcinidae	*Johngarthia weileri*	Cracke
Crustacea	Grapsidae	*Goniopsis pelii*	Caranguejo preto
Crustacea	Grapsidae	*Grapsus adscensionis*	Rock crab
Crustacea	Ocypodidae	*Afruca tangeri*	Fiddler caranguejo
Crustacea	Ocypodidae	*Ocypode cursor*	Phantom caranguejo
Crustacea	Palaemonidae	*Palaemon concinnus*	Mangue shrimp
Crustacea	Portunidae	*Callinectes sp.*	Tartaruga caranguejo
Crustacea	Sesarmidae	*Perisesarma guttatum*	
Crustacea		*Crab sp 2 no ID*	
Mollusca	Archatinidae	*Archachatina bicarinata*	Buzio de terra
Mollusca	Arcidae	*Senilia senilis*	Almexa
Mollusca	Hipponicidae	*Hipponix antiquatus*	
Mollusca	Littorininae	*Echinolittorina tuberculata*	
Mollusca	Littorininae	*Echinolittorina caboverdensis*	
Mollusca	Mactridae	*Leptospisula nívea*	Almexa
Mollusca	Muricidae	*Murex fuscatus*	
Mollusca	Neritidae	*Clypeolum owenianum*	
Mollusca	Ostreidae	*Saccostrea cucullata*	Oyster
Mollusca	Solecurtidae	*Tagelus adansonii*	Almexa
Mollusca		*Gastropoda sp.no ID*	
Mollusca		*Ostreidae sp. No ID*	Oyster

Table 8.4 List of fish species registered in Praia Salgada

Family	Species	Common name	Phase
Bothidae	*Arnoglossus imperialis*	Sole	J
Elopidae	*Elops senegalensis*		J
Gerreidae	*Eucinostomus melanopterus*		J/A
Gobiidae	*Periophthalmus barbarus*	Cucumba	J/A
Gobiidae		Xarroco sp. 1	J/A
Gobiidae		Xarroco sp. 2	J/A
Haemulidae	*Plectorhinchus macrolepis*	Peixe porco	J
Haemulidae	*Pomadasys sp.*	Snorer	J
Lutjanidae	*Lutjanus agennes*	Sea bass	J
Lutjanidae	*Lutjanus sp. 1*	Sea bass	J
Mugilidae	*Liza dumerili*	Tainha	J
Mugilidae	*Liza grandisquamis*	Tainha	J
Poeciliidae	*Aplocheilichthys spilauchen*	Cough-cough	J/A

8.3.5 Physicochemical Parameters

The winding Agua Grande River feeds the Praia Salgada mangrove swamp (with little flow during the Gravana season and considerably more during the rainy season). It flows into the small, sheltered bay of Abade, protected from the wind and the onslaught of the northbound sea. The river flows into the eastern part of Praia Salgada in a narrow inlet about 20 m wide, so the tidal action in this area is evident. At incoming tide, small trains of waves go up to the lagoon in the transition zone. This area extends several hundred metres into a small grassy marsh where the last individuals of *Rhizophora harrisonii can be* found (Fig. 8.3).

The constant input of fresh water and the tides allow for a high renewal rate of the water bodies within the estuary. Regarding the salinity profiles, two different water masses can be observed: salt water and fresh water. However, the low value of seawater in the mouth area (less than 30 ppm) stands out. This is probably due to the contribution of fresh water that has been occurring over time and which mixes to decrease the salinity of the surrounding seawater mass. As this mass of seawater enters the transition zone, the mixing process tends to equalise the parameters, as seen in this case in the salinity. In the inner area, where the tidal influx does not reach, and the freshwater input is higher, the salinity value logically decreases to brackish water values (15–20 ppm).

The oxygen concentration graph shows how the mass of seawater has a typical value (around 80 mg/l), which slightly decreases as it enters the estuary, possibly due to the oxidation process of the organic matter deposited on the bottom. The freshwater mass increases considerably in the transition zone and the mouth of the estuary due to the mixing process caused by marine hydrodynamics.

As the water flows down the transition zone, the depth and the channel increase, so the hydrodynamic decreases, and the organic particles settle to the bottom and increase their value (very high values of more than 160 NTU at some points in the transition zone).

The temperature and pH values remain within the normal range (between 29 and 30 °C; pH close to 8, alkaline) and tend to equalise due to the interaction between the two water masses. Thus the lower temperature of the freshwater increases slightly to match that of the seawater, and the lower pH of the freshwater tends to decrease that of the saltwater as it enters the estuary up to the head of the river.

8.4 Restoration and Ecotourism Activities

Within the scope of the UNESCO Biosphere Reserve activities, the results of the biophysical characterisation and conservation status of the mangrove of Praia Salgada were shared with the community to jointly analyse the possibilities of promoting a restoration action and the development of ecotourism activities.

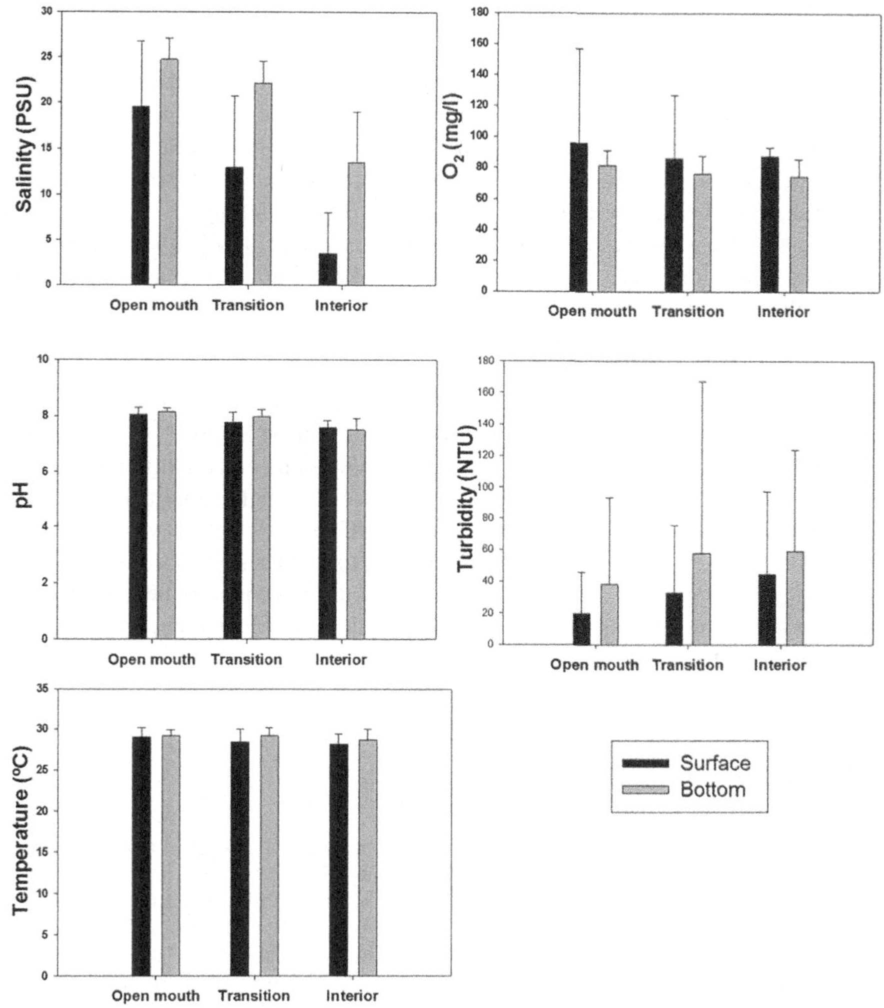

Fig. 8.3 Graphical representations of physicochemical parameters of Praia Salgada

This participatory process resulted in the suggestion of installing a trail to be implemented in the mangrove area, including a birdwatching post. As a complement, the Biosphere Reserve offered training in ecotourism, bird watching, and bird monitoring, activities to be carried out by community members.

Regarding the ecological restoration of the mangrove, the Biosphere Reserve technicians and community members were trained in nursery assembly techniques, reproduction and planting of mangroves, with the support of the SMILO programme—Small Sustainable Islands, through which more than 1100 new young mangrove plants have been planted during the first year of intervention.

8.5 Conclusion

Although preliminary, the biological and physicochemical characterisation of the mangrove of Praia Salgada allowed establishing a reference situation, identifying an opportunity to involve the community in the ecological restoration of this important habitat, associating the conservation component with the development of nature tourism activities with direct benefit for the local microeconomy.

Ecotourism has been shown to offer socio-economic utility for territories by fostering education and social appreciation of biocultural and natural diversity among the inhabitants, generating jobs and income for the populations living in nature reserve areas (Pinkus-Rendón and Pinkus-Rendón 2015).

Several studies confirm that once ecotourism services are established, people employed in this activity tend to become aware of the importance of natural resource conservation and, therefore, of the preservation of natural heritage, which favours sustainability in the reserve and community development (Pinkus-Rendón and Pinkus-Rendón 2015). Thus, this modality reverses the high social and ecological costs traditionally attributed to tourism (Bringas Rábago 1997).

In the case of Praia Salgada in Príncipe Island, tourism activities such as the trail, bird and plant watching will provide, in addition to the economic benefit, relevant information for monitoring the progress of the mangrove plantation and restoration and, consequently, the associated biodiversity and ecosystem services.

References

Bringas Rábago NL (1997) Las dos caras del turismo: Beneficios económicos contra costos socio-culturales y ecológicos: El caso de México. Fermentum 18:89–106

IUCN (2017) International union for the conservation of nature 2017. Red List of Threatened Species 2017:1. http://www.iucnredlist.org

Melo M, Jones P (2008) Bird speciation in the Gulf of Guinea Island system

Pinkus-Rendón MJ, Pinkus-Rendón MÁ (2015) El ecoturismo: quimera o realidad de desarrollo en la Reserva de la Biosfera Ría Celestún. México LiminaR 13(1):69–80

Vargas G (2011) Botánica general: desde los musgos hasta los árboles. EUNED. ISBN:978-9968-31-771-9

Chapter 9
Mangroves of the Brazilian Coastal Amazon: Preservation and Threats

Luis E. Aragón and Miguel Clüsener-Godt

Abstract Mangroves of the Amazon are considered among the most preserved in the world, but regardless of their importance as relevant blue carbon ecosystems, they have not deserved much attention. This chapter presents the situation of mangroves of the Brazilian Coastal Amazon, discusses the main driven factors for its preservation and potential threats, and proposes some recommendations for sustaining mangroves preservation in the area. Among the drivers of preservation are discussed: low human population, legal protection, small-scale economic activities, and lack of large infrastructural complexes. Among the threats are discussed: urban development, road development, unsustainable practices, and oil discovery along the Amazonian coast.

Keywords Mangroves · Brazil · Amazonian coast · Blue carbon · Preservation · Threats

9.1 Introduction

The Amazon reached the most important scientific, political, and economic fora of the world, especially due to its relevant role in climate change and its enormous deposits of natural resources. In fact, the Amazon attracts the world attention as the

L. E. Aragón (✉)
Center for Advanced Amazonian Studies of the Federal University of Pará, Belém, Brazil

UNESCO Chair on South-South Cooperation for Sustainable Development, Federal University of Pará, Belém, Brazil
e-mail: aragon_naea@ufpa.br

M. Clüsener-Godt
UNESCO Chair on Biodiversity Safeguard for Sustainable Development, CFE, Centre for Functional Ecology – Science for People & the Planet, Department of Life Sciences, University of Coimbra, Coimbra, Portugal
e-mail: clusenergodt@uc.pt

M. Clüsener-Godt et al. (eds.), *Blue Carbon Mangrove Ecosystems*, Blue Carbon Ecosystems for Sustainable Development 1,
https://doi.org/10.1007/978-3-031-69553-7_9

largest humid tropical forest on the planet; the largest watershed on Earth; the largest biodiversity reserve and gene bank in the world; and one of the richest mineral provinces that exist, among other superlatives (Mello 2015). Still, despite the progress made, the mangroves of the Amazon, regardless of their importance as relevant blue carbon ecosystems, have not deserved much attention.

Within that context, and using diverse bibliographic sources, this chapter briefly describes the Amazon as an international region, presents the situation of mangroves of the Brazilian Coastal Amazon, discusses the main driven factors for its preservation and potential threats, and proposes some recommendations for sustaining mangroves preservation in the area.

9.2 The International Dimension of the Amazon

The Amazon is an international region of about 7 and a half million km^2 shared by nine countries with varying proportions: Brazil (67.7%), Peru (10.4%), Colombia (6.4%), Bolivia (5.3%), Guyana (2.9%), Venezuela (2.4%), Surinam (2.2%), Ecuador (1.6%), and French Guyana (French Territory) (1.1%) (Aragón 2013) (Fig. 9.1). Based on the population censuses of those countries, the total human population of that immense region can be estimated between 38 and 40 million: Brazil (73.7%), Peru (13.3%), Colombia (3.3%), Bolivia (2.6%), Ecuador (2.5%), Guyana (2.0%),

Fig. 9.1 The Amazon Region. Elaborated by Jhon Williams Montoya Garay (2023). Published with authorization of the author

Surinam (1.4%), French Guyana (0.8%), and Venezuela (0.4%); and most of the population lives in urban areas, including cities of more than one million people.

The great diversity in biological, cultural and landscape terms of this huge region is, therefore, not surprising. Certainly, it is not a homogeneous region; although humid tropical forests predominate, there are also dry tropical forests, flood forests, prairies, agriculture, and other types of vegetation and landscapes (Eva and Huber 2005).

The international dimension of the Amazon at the regional level reveals, therefore, an enormous complexity in physical and human terms. Some of the region's landscapes only exist in certain parts and their degradation can exert serious socio-environmental impacts on other parts of the region, such as mangroves along the coast.

On the other hand, at the global level, as the enormous stock of natural resources in the Amazon demanded by the world economy and the crucial role that the region plays in global climate change are better known, worldwide attention on the region's fate is increasing (Aragón 2018). The continuous economic valorization of nature has turned the Amazon into a global free commercial brand that symbolizes environment, ecology, Indians, traditional populations, forest, oxygen, water, biodiversity, preservation, etc., and in this sense, adding the term Amazon (or something that refers to it) to products and services increases their economic value (Amaral Filho 2016).

9.3 Mangroves of the Brazilian Coastal Amazon

The Amazon River is recognized as the longest and widest in the world. With approximately 7,000 km long, its hydrographic system covers the most extensive basin on the planet, its width varies with the rainy regime, reaching up to 50 km during the rainy seasons, at its delta it expands 320 km width, and the discharge of its waters into the Ocean is estimated in 220,000m^3 per second (UNEP/ACTO 2009). The force of the waters discharged by the Amazon into the Ocean makes these waters penetrate the Ocean for about 100–150 km from the river-mouth channels (Nittrouer et al. 2021). Also, it is estimated that the Amazon River brings 4.36×10^8 tons per year of suspended sediments into the Ocean (Fagundes et al. 2020), which are spread along the coast by ocean currents up to the Orinoco Delta and the Caribbean Sea (Nittrouer et al. 2021). The Amazon, therefore, determines, in many ways, the functioning of the coastal ecosystems.

Consequently, the shoreline of the Amazon Region as a whole, can be considered as the one from the Delta of the Orinoco River in Venezuela to the San José Bay in the State of Maranhão (Brazil), covering coastal areas of Venezuela, Guyana, Surinam, French Guyana and Brazil. Physical conditions of this extensive area are favorable to the proliferation of mangroves, among others: high temperature, high humidity, extensive flooding area, and principally, enormous volumes of fresh water and amounts of sediments supplied to the sea by the Amazon River (Spalding et al. 2010; Hayashi et al. 2023; Nittrouer et al. 2021).

In Brazil, the second country in the world with most extensive areas covered with mangroves, after Indonesia, estimated by Spalding et al (2010) in 13,000 km^2, mangroves cover extensive areas of the coast. More recent figures calculated in the basis of satellite data analysis estimated in some 9900 km^2 the total area covered by mangroves in Brazil in 2018, most of which (7280 km^2) located in the coastal areas of the states of Amapá (830 km^2), Pará (2100 km^2) and Maranhão (4350 km^2) (Diniz et al. 2019). The coastal area that goes from São José Bay in Maranhão to Marajó Bay in Pará, known as the Amazon mangroves belt, is considered the most extensive, dense, continuous, and preserved in the world (Hayashi et al. 2019, 2023).

Mangroves are unique ecosystems with the dual capacity of being at the same time a sink for atmospheric CO_2 and essential source of oceanic carbon (Saint-Paul and Schneider 2010). In the Amazon, blue carbon stocks of mangroves exceed those of upland tropical evergreen forests ecosystems (Kauffman et al. 2018),[1] and those of other coastal areas of Brazil (Hatje et al. 2020).[2] Therefore, mangroves play an important role in mitigating climate change, and their deforestation "with their extraordinary high rates of primary productivity drastically reduces these capacities" (Saint-Paul and Schneider 2010).

According to Hatje et al. (2020), mangroves play many roles and provide multiple environmental services, because, among other things, they control exchanges of materials at the interfaces between land, atmosphere, and ocean ecosystems, they protect the coasts by controlling erosion, they protect communities living in coastal areas of storms and tsunamis, and they contribute to food security providing food resources. In the Brazilian Coastal Amazon, mangroves have significant socio-economic importance for the local traditional communities, especially for fisheries, capture of shrimps, crabs, and other species, as well as a source of lumber for construction and production of charcoal to supply energy (Hayashi et al. 2019).

[1]"The mean carbon stocks of the Amazon mangroves were over twice those of upland Amazon evergreen forests, exceeding their mean by about 245 Mg C ha^{-1}" (Kauffman et al. 2018).

[2]"The low carbon stocks of the Northern Brazil mangroves soils according to Kauffman et al. (2018) could be explained by the Amazon's region large tidal range, which does not favor the establishment of anoxic soil conditions, added to the coarse nature of soils, which is not as efficient to trap organic C as the fine textured soils in the Jaguaripe, for instance. However, when comparing mean ecosystem carbon stocks per unit of area, the Amazonian mangroves (361–746 Mg C ha^{-1}; (Kauffman et al. 2018)) are of the same magnitude as the Jaguaripe region (250–633 Mg C ha^{-1}), clearly indicating that above ground stocks for ecosystems estimates are more important in the Amazonian than in the Eastern ecoregion" (Hatje et al. 2020).

9.3.1 Preservation

According to several sources, worldwide, mangroves have been shrinking due to human activities. Between 2000 and 2012 it is estimated that mangrove habitats were reduced at a rate of 0.39% per annum around the world; still, and regardless of anthropogenic pressures, mangroves of the Brazilian Coastal Amazon remain still relatively well preserved: anthropogenic activities have impacted less than 1% (Hayashi et al. 2023). Several factors contribute to this preservation including the following:

1. Low population. Figure 9.2 illustrates the distribution of total population and mangrove areas in 2022 (IBGE 2023; MapBiomas 2023) over the grid of municipalities legally included in the list of the terrestrial coastal zone of the states of Amapá (AP), Pará (PA), and Maranhão (MA).[3] Total area of the

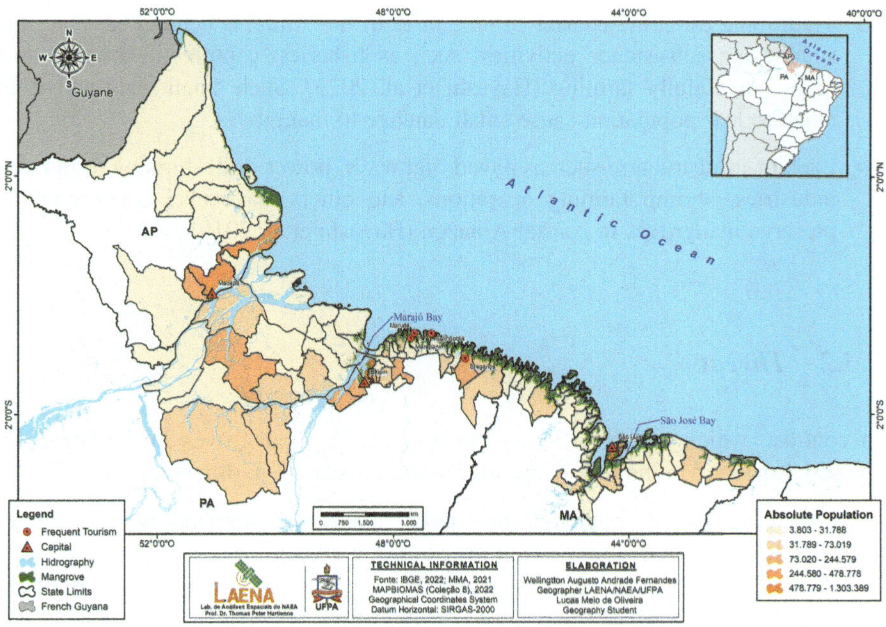

Fig. 9.2 Mangrove areas and total population of the municipalities of the terrestrial coastal zone of the states of Amapá (AP), Pará (PA) and Maranhão (MA), 2022. Elaborated by Wellingtton Augusto Andrade Fernandes and Lucas Melo de Oliveira (2023). Published with authorization of the authors

[3] The coastal zone in Brazil is defined by law: "The Brazilian coastal zone, considered national heritage by the 1988 Constitution, corresponds to the geographic space of interaction of air, sea and land, including its renewable or non-renewable resources, covering a maritime strip and a land strip, with the following limits: I - maritime strip: space that extends for 12 nautical miles, measured from the baselines, thus comprising the entirety of the territorial sea; II - terrestrial strip: space comprised

98 municipalities of that zone sums 252,234 km^2, with a total human population of 6,677,306 (26.47 people per km^2) unevenly distributed, varying from 3803 in Pracauúba (Amapá) and 1,303,389 in Belém (capital of the state of Pará); the other two capitals reach 1,037,775 people in São Luis (Maranhão) and 442,933 in Macapá (Amapá) (IBGE 2023). Notice, however, that not all those municipalities have mangroves in their territories (MapBiomas 2023)

2. In Brazil, mangroves are protected by law. The Brazilian Forest Code characterizes mangroves as areas of permanent preservation (Brasil 2012), and the establishment of 18 protected areas of different types in the Brazilian Amazon coast, have legally assured the conservation of the ecosystem and its biodiversity (Hayashi et al. 2023). According to ICMBio (2018) those protected areas have also contributed to provide the conditions to maintain the socio-economic and cultural needs of the traditional estuarine-coastal communities, and the protection of the goods and services provided by mangroves (ICMBio 2018; Hayashi et al. 2023).
3. Traditional estuarine-coastal communities of the Amazon depend basically on small-scale subsistence activities, such as fisheries, capture of shrimps and crabs, and family farming (Hayashi et al. 2023). Such small-scale activities allied to low population cause small damage to mangroves.
4. Lack of infrastructure such as paved highways, power grid, shipping terminals, industries, shrimp farming operations, and others, have also contributed to preserve mangroves in coastal Amazon (Hayashi et al. 2019).

9.3.2 *Threats*

In contrast, which are the potential threats faced by mangroves in the Brazilian Coastal Amazon? In the following, four major threats are highlighted:

1. Urban development. Although the coastal zone is characterized by low population, the population is not evenly distributed, presenting large unoccupied areas interspersed with places of high population and some larger cities in continuous urban expansion (Szlafsztein 2009), which often result in draining mangrove areas. In the same line of analysis, possible expansion of infrastructure for tourism, such as hotels and resorts, is also a potential threat for mangroves especially in those areas of beautiful beaches such as the areas around Salinópolis, Marudá, Bragança, Marapanim, and others.

by the limits of the Municipalities that are directly influenced by the phenomena occurring in the coastal zone" (Brasil, 2024—Decreto 5.300, 7/12/2004, art. 3). Updated list of municipalities of the terrestrial strip of the states of Amapá, Pará, and Maranhão includes 98 municipalities (Brasil 2021—Portaria MMA no. 34, 03/02/2021).

2. Road development. The Amazon mangroves belt area is served by a network of some 19,520 km of roads, 90% unpaved, which have contributed to preserve mangroves (Hayashi et al. 2019). However, as those authors point out, pressure for paving of roads and construction of others makes road development as a potential threat, because, among other impacts, it generates changes in the hydrologycal regime as happened with paving of the state road PA-158 that links Bragança with the beach of Ajuruteua (26 km), that destroyed 6 km^2 of mangroves when interrupting the flow of sea water to the mangroves area (Hayashi et al. 2019).
3. Unsustainable practices. Mangroves of the Brazilian Coastal Amazon are ecosystems of high biodiversity and rich in natural resources that have been exploited so far only in small scale with little damage. Still some authors alerted about the possibility of enlargement of extractive activities of the fauna and deforestation of mangroves for construction and generation of energy (Hayashi et al. 2019), which will be very harmful for mangroves.
4. Oil Discovery. News around the world inform about the discovery of enormous reserves of petroleum in the sea platform of Guyana and Surinam that overpasses the 10 billion barrels and is expected that their exploitation will radically change the landscape of the coasts of those countries; and it is suspected that areas of the reserves expand to the whole Amazonian coast platform. In Brazil, intense debate exists concerning the prospection of oil reserves in the coast of the state of Amapá. The exploitation of oil on a large scale in those areas will certainly be a potential threat for conservation of mangroves and survival of local communities in the Brazilian Amazon Coast. Furthermore, in the case of oil spills, they would be transported along the whole Amazonian coast, an issue that would certainly take an international dimension (Pires do Rio 2023). It is important to remember that French Guyana is an European territory, who counts with rigid environmental laws.

9.4 Conclusions and Recommendations

9.4.1 Conclusions

The studies revised in this chapter reveal that mangroves of the Brazilian coastal Amazon are among the most preserved of the world. The question is: for how long?

Through time, local conditions have protected those ecosystems of depredation, including as main drivers, low population, lack of large infrastructures such as highways, industries, large ports, small-scale family agriculture and fisheries, domestic low impact tourism, and the protection of mangroves by law, among others. Still, in the globalized world of today, where the Amazon gains world importance, the mangroves of the Brazilian Amazon Coast face potential threats, as is happening in other parts of the world. Pressure of urban development, road improvement and construction, massive tourism, and expansion of unsustainable

Table 9.1 The ten municipalities with the most extensive area of mangroves in Brazil

Municipality	State	Area of mangroves (hectares) 2020[1]	Human population 2022[2]	HDI 2010[3]
1 Amapá	Amapá	56,713	7943	0.642
2 Cururupu	Maranhão	50,902	31,558	0.612
3 Viseu	Pará	39,841	58,692	0.515
4 Augusto Corrêa	Pará	30,014	44,573	0.520
5 Turiaçu	Maranhão	29,855	37,491	0.561
6 Humberto de Campos	Maranhão	29,495	25,669	0.535
7 Candido Mendes	Maranhão	28,953	19,932	0.561
8 Cajapió	Maranhão	26,863	10,121	0.553
9 Carutapera	Maranhão	25,683	24,238	0.574
10 Godofredo Viana	Maranhão	25,131	10,186	0.604
Total		317,767	270,403	

Sources: 1. MapBiomas (2021), 2. IBGE (2023), 3. UNDP et al. (2023)

exploitation of mangrove resources are threats that if not controlled will hamper mangroves preservation in shorter time. Furthermore, discovery and exploitation of oil along the Brazilian Amazon coast platform, as already happens in Guyana and Surinam, would have tremendous negative impacts on mangroves if innovative measures are not taken for its exploitation in order to have a low impact on those areas that have a rich biodiversity. It is of utmost importance to preserve those fragile ecosystems including their relevant socio-environmental services. Some argue that other sources of cleaner energy should be sought in these times of energy transition (Pires do Rio 2023).

In that perspective, it is important to consider that despite its natural wealth, the Amazon Region registers some of the lowest Human Development Indexes (HDI) of the country, including all the miseries of underdevelopment, which demands challenging measures in search of the Sustainable Development Goals defined by the United Nations (Aragón 2015). In fact, the states of Para, Amapá and Maranhão occupy the 23rd, 25th and 27th places in the ranking of the 27 Federal Units of Brazil with HDI of 0.690, 0.688 and 0.676 respectively in 2021 (UNDP et al. 2023). At the municipal level, the ten municipalities with more extensive mangrove areas in Brazil, which are all located in the Amazon Coast, cover a total area of 317,767 hectares of mangroves (MapaBiomas 2021), and the HDI varies from 0.642 in Amapá (Amapá) to 0.515 in Viseu (Pará) (Table 9.1).

9.4.2 Recommendations

Given the scenario of preservation of mangroves in the Brazilian Coastal Amazon and the identification of the main drivers of such preservation and potential threats, it is relevant to mention some actions that could contribute to sustain such preservation within the principles of sustainable development.

Existing capacity in the area can contribute significantly to improve scientific knowledge of mangroves in the Amazonian coast. Universities and research institutes located in the capitals, Belém, Macapá, and São Luis, some of them with research and graduate programs on topics related to mangroves could cooperate to maximize human and physical resources to investigate mangroves. Furthermore, some of those universities have already *campi* and laboratories in the cities of the coast.

Further studies should be conducted, among other purposes: (1) to better know the flora and fauna of mangroves of the Amazonian coast, (2) to determine more accurately their socio-environmental services, (3) to better identify the potential threats of mangroves of the Amazonian coast, including, oil exploitation, (4) to better understand the spatial distribution of anthropogenic activities in the mangroves and their socio-environmental impacts and asses the effectiveness of conservation units in the region (Hayashi et al. 2023), (5) to better understand mangroves West of Marajó Bay. Scientific literature on mangroves of the Brazilian Coastal Amazon has been concentrated on the mangroves belt (East of Marajó Bay). West of Marajó Bay, mangroves have not received the same attention. There are evidences that mangroves from this part of the Brazilian coast are somehow different than the ones located East of that Bay. Because of a stronger influence of the huge volume of water and sediments brought to the Ocean by the Amazon River, mangroves of this area have proliferated in fresh water or with very low levels of salinity (Bernardino et al. 2022; Visschers et al. 2022). This fact highlights the strong potential of mangroves to adapt to changing environments and different levels of salinity.

The UNESCO Chair on South-South Cooperation for Sustainable Development can be an important instrument to strengthen scientific cooperation between local and foreign institutions. That UNESCO Chair was established at the Center for Advanced Amazonian Studies of the Federal University of Pará in Belém (NAEA/UFPA), Brazil in 2006, with the following main objectives: (1) To develop scientific knowledge and to improve South-South cooperation for sustainable development thus helping to achieve concrete results in order to improve synergy in the delivery of high-quality scientific information on the current understanding of environmental change, emerging environmental issues, the interactions between humankind and the environment, and their consequences for human well-being; (2) to facilitate collaboration between high-level, internationally recognized researchers and teaching staff of the University and other institutions in Brazil and elsewhere in Latin America and the Caribbean, and in other regions of the world, (3) to implement a broad system of activities and partnerships related to higher education, research and documentation, and in particular to major themes such as the population and the environment in the Amazon region and the Biosphere Reserves Program. Within those objectives, the UNESCO Chair on South-South Cooperation for Sustainable Development can establish linkages with other UNESCO Chairs focusing on blue carbon ecosystems such as the UNESCO Chair on Education in Biosphere Reserves for Sustainable Societies of Yokohama National University of Japan and the UNESCO Chair on

Biodiversity Safeguard for Sustainable Development of the University of Coimbra in Portugal.

Finally, the establishment of Biosphere Reserves in the Amazonian Coast would be another asset for the sustainable use of mangroves. Frequently, the implementation of protected areas tends to focus on the values of landscapes in ecological terms overseeing the pressing livelihood needs of local populations. UNESCO Biosphere Reserves consider the specific settings selected by the countries, which integrate management strategies with the participation of different stakeholders including the local people, the private sector, the government, NGOs, and the scientific community, seeking to reconcile conservation of biological and cultural diversity and economic and social development through partnerships between people and nature (Aragón and Clüsener-Godt 2013). Efforts already made through the Chair on South-South Cooperation for Sustainable Development to establish Biosphere Reserves in the coasts of the State of Pará and of Surinam, should further continue.

Acknowledgments Thanks to the Yokohama National University of Japan for the support given to Luis Aragón to participate of the international seminar in January 2023 where this paper was presented. Also, thanks to Prof. Jhon Williams Montoya Garay of the Department of Geography of the National University of Colombia for elaborating Fig. 9.1 and to Wellingtton Augusto Andrade Fernandes and Lucas Melo de Oliveira of the Laboratory of Spatial Analysis of the Center for Advanced Amazonian Studies of the Federal University of Pará for elaborating Fig. 9.2. This study was partially funded by a scholarship awarded to Luis Aragón by the National Council of Scientific and Technological Development of Brazil (CNPq).

References

Amaral Filho O (2016) Marca Amazônia: o marketing da floresta. CRV Editora, Curitiba. https://www.editoracrv.com.br/produtos/detalhes/32242-marca-amazoniabr-o-marketing-da-floresta

Aragón LE (2013) Amazônia: Conhecer para desenvolver e conservar, cinco temas para um debate. Hucitec, São Paulo. https://lojahucitec.com.br/produto/amazonia/

Aragón LE (2015) Desenvolvimento amazônico em questão. Revista Crítica de Ciências Sociais 107:5–16. https://doi.org/10.4000/rccs.5983

Aragón LE (2018) A dimensão internacional da Amazônia: um aporte para sua interpretação. Revista NERA 21(42):14–33. https://revista.fct.unesp.br/index.php/nera/issue/view/28%20de%20mar.%202018

Aragón LE, Clüsener-Godt M (2013) Biosphere Reserves and UNESCO Chairs: partnerships for sustainable development. Papers do NAEA 315:1–27. https://periodicos.ufpa.br/index.php/pnaea/article/view/11302/7783

Bernardino AF, Mazzuco ACA, Souza FM, Nobrega GN, Silva TSF, Kauffman JB (2022) The novel mangrove environment and composition of the Amazon delta. Curr Biol 32:3636–3640. https://doi.org/10.1016/j.cub.2022.06.071

Brasil (2004) Decreto no. 5.300 de 7 de dezembro de 2004. https://www.planalto.gov.br/ccivil_03/_ato2004-2006/2004/decreto/d5300.htm

Brasil (2012) Lei no. 12.651, de 25 de maio de 2012. https://www.planalto.gov.br/ccivil_03/_ato2011-2014/2012/lei/l12651.htm

Brasil (2021) Portaria do Ministério de Meio Ambiente MMA no. 34, de 03 de fevereiro de 2021. chrome-extension://efaidnbmnnnibpcajpcglclefindmkaj/https://www.sema.ce.gov.br/wp-content/uploads/sites/36/2021/02/PORTARIA-MMA-No-34-DE-2-DE-FEVEREIRO-DE-2021-PORTARIA-MMA-No-34-DE-2-DE-FEVEREIRO-DE-2021-DOU-Imprensa-Nacional.pdf

Diniz C, Cortinhas L, Nerino G, Rodrigues J, Sadeck L, Adami M, Souza-filho PWM (2019) Brazilian mangrove status: three decades of satellite data analysis. Remote Sens 11(808):1–19. https://doi.org/10.3390/rs11070808

Eva H, Huber O (eds) (2005) A proposal for defining the geographical boundaries of Amazonia. European Commission, Joint Research Centre, Publication Office, Luxembourg. https://op.europa.eu/en/publication-detail/-/publication/98b46d75-e3e7-4208-a104-465bf28e6703

Fagundes HO, Fan FM, Paiva RCD, Siqueira VA, Buarque DC, Kornowski LW, Laipelt L, Collischonn W (2020) Sediment flows in South America supported by Daily Hydrologic-Hydrodynamic Modeling. Water Resour Res 57/2. https://doi.org/10.1029/2020WR027884

Hatje V, Masqué P, Patire VF, Dórea A, Barros F (2020) Blue carbon stocks, accumulation rates, and associated spatial variability in Brazilian mangroves. Limnol Oceanogr 999:1–14. https://doi.org/10.1002/lno.11607

Hayashi SN, Souza-Filho PWM, Nascimento WR Jr, Fernandes MEB (2019) The effect of anthropogenic drivers on spatial patterns of mangrove land use on the Amazon coast. PLoS One 14:1–20. https://doi.org/10.1371/journal.pone.0217754

Hayashi SN, Souza-Filho PWM, Nascimento WR Jr, Fernandes MEB (2023) Status of mangroves land use on the Brazilian Amazon coast from RapidEye imagery and GEOBIA approach. An Acad Bras Cienc 95/2. https://www.scielo.br/j/aabc/a/vcswV7jTXn5ZpRWrsmDLdMn/?lang=en

IBGE – Instituto Brasileiro de Geografia e Estatística (2023) Censo 2022. https://censo2022.ibge.gov.br/

ICMBio – Instituto Chico Mendes de Conservação da Biodiversidade (2018) Atlas dos manguezais do Brasil. Instituto Chico Mendes de Conservação da Biodiversidade, Brasília. https://www.gov.br/icmbio/pt-br/centrais-de-conteudo/atlas-dos-manguezais-do-brasil-pdf/view

Kauffman JB, Bernardino AF, Ferreira TO, Giovanni LR, Gomes LEO, Romero DJ, Jimenez LCZ, Ruiz F (2018) Carbon stocks of mangroves and salt marshes of the Amazon Region, Brazil. Biol Lett 14(20180208):1–4. https://doi.org/10.1098/rsbl.2018.0208

MapBiomas (2021) Manguezais, Praias, Dunas & Areais, Aquicultura e Apicuns nos Últimos 36 Anos: Destaques do mapeamento anual dos manguezais, praias, dunas, areais, aquicultura e apicuns no Brasil entre 1985 a 2020. MapBiomas – Mapeamento Anual de Cobertura e Uso da Terra do Brasil Coleção 6. https://brasil.mapbiomas.org/wp-content/uploads/sites/4/2023/09/MapBiomas_Zona_Costeira_Outubro_2021_30102021_OK.pdf

MapBiomas (2023) Download direto da Coleção 8 com mapas de Cobertura e Uso da Terra de todo o Brasil em formato GeoTiff (um mapa para cada ano). Para o ano 2022: https://storage.googleapis.com/mapbiomas-public/initiatives/brasil/collection_8/lclu/coverage/brasil_coverage_2022.tif

Mello AF (2015) Dilemas e desafios do desenvolvimento da Amazônia: o caso brasileiro. Revista Crítica de Ciências Sociais 107:91–108. https://doi.org/10.4000/rccs.6025

Nittrouer CA, DeMaster DJ, Kuehl SA, Figueiredo AG Jr, Sternberg RW, Faria LEC, Silveira OM, Allison MA, Kineke GC, Ogston AS, Souza Filho PWM, Asp NE, Nowacki DJ, Fricke AT (2021) Amazon sediment transport and accumulation along the continuum of mixed fluvial and marine processes. Annu Rev Mar Sci 13:501–536. https://www.annualreviews.org/doi/10.1146/annurev-marine-010816-060457

Pires do Rio GA (2023) Margem equatorial brasileira: desafios postos sobre a mesa. Revista Brasileira de Energia 29(1):88–105. https://doi.org/10.47168/rbe.v29i1.742. https://sbpe.org.br/index.php/rbe/article/view/742/554

Saint-Paul U, Schneider H (2010) Epilog. In: Saint-Paul U, Schneider H (eds) Mangrove Dynamics and Management in North Brazil. Springer, Berlin, pp 389–393. https://link.springer.com/book/10.1007/978-3-642-13457-9

Spalding M, Kainuma M, Collins L (2010) World Atlas of Mangroves. Earthscan, London \Washington, D.C.. https://books.google.com.br/books?id=wzSCkulW9SQC&printsec=frontcover&hl=pt-BR&source=gbs_ge_summary_r&cad=0#v=onepage&q&f=false
Szlafsztein CF (2009) Indefinições e obstáculos no gerenciamento da Zona Costeira do Estado do Pará, Brasil. Revista de Gestão Costeira Integrada 9:47–58. https://doi.org/10.5894/rgci114
UNDP – United Nations Development Programme/FJP – Fundação João Pinheiro/IPEA – Instituto de Pesquisa Econômica Aplicada (2023) Atlas do Desenvolvimento Humano no Brasil. http://www.atlasbrasil.org.br/ranking
UNEP/ACTO – United Nations Environmental Programme/Amazon Cooperation Treaty Organization (2009) Geoamazonia: Environment outlook in Amazonia. UNEP/ACTO, Brasília. https://wedocs.unep.org/20.500.11822/9421
Visschers LLB, Santos CD, Franco AMA (2022) Accelerated migration of mangroves indicate large-scale saltwater intrusion in the Amazon Coastal wetlands. Sci Total Environ 800:1–8. https://www.sciencedirect.com/science/article/pii/S0048969722027759?via%3Dihub

Chapter 10
The Man and the Biosphere (MaB) Programme's Role Towards the Effective Management of Blue Carbon Ecosystems

Benno Böer and Neha Midha

Abstract Climate change has become a cross-cutting global reality, affecting humanity's energy-, food-, and water-security. Therefore, investigating options towards climate adaptation, mitigation, and resilience are urgently required. UNESCO's World Network of Biosphere Reserves already plays several very important roles, including awareness-raising, increasing the level of community-shock-resielience through improved knowledge and skills, via science-based environmental education. Moreover, coastal Biosphere Reserves, with their seagrass, salt marsh and mangrove ecosystems are known to sequester large amounts of atmospheric carbon, and lock it away for very long time-periods, in their semi-sub-hydric soils. This contribution discusses UNESCO Biosphere Reserves in the context of blue carbon ecosystem management, and the contemporary and future roles of Biosphere Reserves.

Keywords Biosphere reserves · Conservation · Restoration · Blue carbon ecosystems · Mangroves · Salt marshes · Seagrass beds

10.1 Background & Introduction

World Wars I. and II. were so immensely devastating and caused so much human loss, tragedies and pain, that nations decided to form the United Nations which was mandated to minimize conflicts including wars, and to foster peace. In the same year (1945) UNESCO was formed to assist fostering peace on the platforms of education, science and culture, because political agreements alone are not enough to build long-lasting peace. It was understood and agreed that the intellectual and moral solidarity between nations and people is an absolutely essential element for peace.

B. Böer (✉) · N. Midha
UNESCO New Delhi Office in South Asia for Bangladesh, Bhutan, India, Nepal, the Maldives, and Sri Lanka, New Delhi, India
e-mail: b.boer@unesco.org; n.midha@unesco.org

M. Clüsener-Godt et al. (eds.), *Blue Carbon Mangrove Ecosystems*, Blue Carbon Ecosystems for Sustainable Development 1,
https://doi.org/10.1007/978-3-031-69553-7_10

UNESCO aims at building peace in the minds of men and women. The Preamble to the Constitution of UNESCO declares: '*Since wars begin in the minds of men and women, it is in the minds of men and women that the defenses of peace must be constructed*'. UNESCO works to create the conditions for dialogue among civilizations, based upon respect for shared values. With multiple current crises ongoing simultaneously, including wars and armed conflicts, the Covid-19 pandemic, climate change, biodiversity loss, pollution, the energy crisis, the importance of peace cannot be under-estimated. Peace is a crucial element for human survivability and concerns all of us. Functioning ecosystems, that will continuously supply humanity, currently exceeding 8 billion people, with air to breath, water to drink, and food to eat, is an absolute essential pre-condition for peace. The Man and the Biosphere Programme (MaB), is one of UNESCO's most important initiatives advocating to live in harmony with nature. MaB supports the identification, establishment and management of Biosphere Reserves. The MaB Programme and the World Network of Biosphere Reserves is of great value for the conservation, restoration and scientific research concerning Blue Carbon Ecosystems (BCE).

The MAB programme is an intergovernmental scientific programme that aims to establish a scientific basis for enhancing the relationship between people and their environments. It combines the natural and social sciences with a view to improving human livelihoods and safeguarding natural and managed ecosystems. It promotes innovative approaches to economic development that are socially and culturally appropriate and environmentally sustainable. BRs are places for conservation, restoration, scientific research, and testing, applying and demonstrating to the international community best practices for humans living in harmony with nature. They aim to enhance nature conservation, reduce environmental footprints, and keep the ecological carrying capacity of natural ecosystems intact. Globally, there are numerous BRs that conserve Blue Carbon Ecosystems, including mangroves.

The scientific community has been using the term 'Blue Carbon Ecosystems' for a little more than 10 years. However, there is no clarity yet on what exactly is a BCE. Seagrass beds, mangroves, salt marshes, and sabkhas in most cases are BCEs (or have been in the geological past). However, we need to know much more about their carbon sequestration capacity which is patchy and diverse. BRs and the MaB Programme can assist raising awareness of the importance of these ecosystems and, via science-education, enhance the knowledge of the general public.

10.2 Japan

The International Conference on Blue Carbon Ecosystems took place in January 2023, in Yokohama, Japan. Japan is an important country for blue carbon aspects for a range of reasons, most importantly because of the decades long work of the International Society for Mangrove Ecosystems (ISME), located in Okinawa. ISME was established in 1990 and took over mangrove restoration projects that were initiated by UNESCO and UNDP. It is also noteworthy to mention the efforts of the Yokohama National University, organizing the Blue Carbon Forum in January 2023. Even though Japan has a total mangrove coverage of only 7.4 km^2, the country is of great bio-geographical interest, because mangroves in Japan are some of the most northerly mangroves occurring across the Indo-West Pacific biogeographical region. Japan is actively participating in the World Network of Biosphere Reserves, with 10 BRs currently listed. Japan has also the economic capacity to provide assistance towards BCE support on a global scale.

10.3 Small Island Development States (SIDS)

The Marshall Islands is one of the Small Island Developing States (SIDS). The Marshall Islands have a total mangrove coverage of only 0.0234 km^2. *Bruguiera gymnorhiza* is known to occur on a number of atoll islands, and other species that have been observed include *Lumnitzera littorea*, *Pemphis acidula*, *Sonneratia alba*, and *Xylocarpus granatum*, as well as *Avicennia* and *Rhizophora* species. It has been argued, that the distribution of *Bruguiera gymnorhiza* on the atolls of some of the Marshall Islands, and the conspicuous absence of these plants on un-inhabited atolls, suggests that human influences might have been at work. Mangrove propagation seems to coincide with actions by the Marshallese to retain land gained during typhoon events. This is of course a highly interesting thought towards nature-based solutions for climate resilience. Mangrove specialist Vainuupo Jungblut (Marshall Islands) confirmed that the mangroves will help protect crops, gardens and communities from storm surges and extreme events thereby increasing community resilience to climate change and sea level rise.

At least 35 of the 39 countries listed as SIDS countries have mangrove ecosystems, and in some of them, the total territory is covered by upto 40%.

10.4 Africa

Africa as a whole has vast areas of blue carbon ecosystems in the east, the southern central parts, and the west. The northern parts, which are too cold for mangrove growth, have vast and rich seagrass beds and salt marshes. Africa has a total of 28,000 KM^2 of mangroves. All coastal countries in Africa have mangroves, with the exception of Namibia (and the Mediterranean). Tanzania, for example, has about 1287 km^2 and Madagascar has 2990 km^2 of mangroves.

10.5 The Arab Region

In the western part of the Arab Region mangroves only occur in Mauretania, and in the eastern part of the Arab Region these ecosystems are restricted to the Red Sea, the Persian Gulf, and the Arabian Sea. Mangroves in Qatar, for example, have a total coverage of 12.3 km^2, and they occur mostly around Al Khor and Al Dhakeera. In the Gulf of Salwa salinity levels exceed 7.0%, a limiting factor for mangrove growth, however, very rich and highly productive seagrass beds occur there. Qatar, just like many other countries globally, has lost quite a bit of mangroves due to coastal development projects, and, at the same time, pioneered successfully an experimental site for floating mangroves in Lusail City, with a survival rate of 95%, which is of great importance for scientific research into future biofuel and carbon sequestration. Many Arab countries have mangroves, even though the Arab Region is characterized by low precipitation rates. This means the utilization of marginal water resources, including seawater, brackish water, and wastewater is of high importance for scientific research into cash crop halophytes for the production of food, fiber, fuel, and other products.

10.6 Asia & The Pacific

The vast region of Asia & the Pacific has most mangroves in terms of area coverage and diversity. Indonesia has the largest total mangrove coverage of any country, with at least 31,890 km^2, slightly more than 20% of the global mangrove forest coverage.

India, for example, has a long coastline of more than 7500 km, and several Biosphere Reserves with BCEs, for example the Gulf of Mannar BR. This BR was established in 2001. This BR is located in Tamil Nadu, India, with its coral reefs, seagrass beds and mangroves, and it is considered one of the world's richest marine ecosystems. The Sundarbans is the largest stand-alone mangrove forest globally, and the largest delta in the world. It extends from Bangladesh (where it is being conserved as a World Heritage Site) into India (where it is also being conserved as a World Heritage Site, as well as a BR, which offers certain advantages, due to the people-centric management requirements). The Indian Sundarban BR was established in 2001.

10.7 Europe & North America

Europe has no mangroves, but vast areas of salt marshes and seagrass beds, macro-algal reefs and other BCEs. The only country in North America with mangroves is the United States of America with a total mangrove forest coverage of at least 3.030 km^2. Mangrove biodiversity and ecosystem dynamics in the USA are among the best scientifically documented in the world, providing an important source of information for ecosystem restoration projects.

10.8 Latin America & the Carribean

Brazil, Mexico and several other countries in Latin America have vast areas of mangroves. After Indonesia, Brazil has more mangrove forests than any other nation, with 13.000 km^2, or 8.5% of the global coverage.

The above figures are based on the World ATLAS of Mangroves (2010).

10.9 Biosphere Reserves: Perspectives for Blue Carbon

BR are places where scientific research should be carried out and methods of atmospheric and oceanic carbon sequestration should be tested and applied. Coastal BRs are often places for blue carbon conservation, restoration, scientific research, offsetting habitat loss, community participation, developing best practices for sustainable ecosystem services, and places for private sector support with Governments and communities.

The MaB Programme can assist to identify, establish and manage new Biosphere Reserves and carry out scientific research and test and apply best practices as solutions for ongoing contemporary crises, including unbiased research into climate change issues, climate resilience, and enhancement of the community's knowledge and skills on climate adaptation.

For Blue Carbon, it is highly important to carry out scientific research, including ground-truthing into the sediments and primary productivity of mangroves, seagrass beds and salt marshes. The obtained data should provide accurate scientific information on BCE's carbon sequestration capacity. It needs a global carbon database, including for ecosystems that are located inside and outside UNESCO-designated sites.

Moreover, it is highly desirable to produce detailed and up-to-date accurate GIS-based maps on the geographical coverage of mangroves, salt-marshes, seagrass beds, delta-systems, coastal brackish deltas and swamps, and compliment them with accurate and reliable layers of sediment-ecology, including sediment depths, and multi-chemical analysis of the sediments, in view of carbon storage and output.

The production of an open access internet-based Blue Carbon ATLAS, coupled with a GIS-based database, would be highly desirable for the accurate updating of the state of global BCEs, and the reliable open-access provision of frequently updated science-based geo-spatial data.

In addition, any effective management of BCEs requires the balanced approach towards conservation and economic development involving communities and the people living inside or close to these areas. The MaB programme promotes and demonstrates innovative, traditional and indigenous knowledge and solutions, as best practices that are socio-ecologically and culturally acceptable.

Finally, the organization of frequent Blue Carbon Forum's with the participation of Government agencies, UN agencies, NGOs, academia, conservationists, restoration specialists, anthropogenic ecosystem designers and practitioners, supported by the private sector, is urgently needed, as an unbiased scientific technology platform that allows participants to exchange views, ideas, and to disagree.

Reference

ISME, ITTO, FAO, UNEP-WCMC, UNESCO, United Nations University & The Nature Conservancy (2010) In: Spalding M, Kainuma M, Collins L (eds) The world mangrove ATLAS. Earthscan Ltd., London, p 319

Chapter 11
Japanese Mangroves as Blue Carbon Ecosystems

Kiyoshi Fujimoto

Abstract In Japan, there are 885 ha of valuable mangrove forests as the northern limit of their distribution in the Pacific Ocean, 604 ha of which are distributed in Iriomote Island in the Yaeyama Islands. The maximum aboveground biomass in Iriomote was estimated to be 361 Mg ha^{-1} for *Bruguiera gymnorrhiza* community. The accretion rate of aboveground biomass is considerably slower than in the tropics, which is largely due to the frequent and intense typhoons in addition to climatic conditions. On the other hand, belowground, *Rhizophora stylosa* communities accumulate mangrove peat and store approximately 500 Mg ha^{-1} of carbon up to 1 m deep, as in the tropics. Although the fine-root production, the main parent material of mangrove peat, is inferior to that in the tropics, it can be evaluated that the belowground plays a significant role as a place of carbon storage.

Keywords Aboveground biomass · Forest dynamics · Belowground stored carbon · Fine roots production · Northern limit mangroves

11.1 Introduction

Japan has valuable mangrove forests as the northern limit of their distribution in the Pacific Ocean. Mangrove forests in Japan are often located on deltas of small rivers (Kikuchi et al. 1978; Ishihara et al. 2004). On the delta, microtopography such as natural levees and back marshes have been formed, and mangrove species are distributed accordingly (Kikuchi et al. 1978; Fujimoto and Ohnuki 1995; Fujimoto and Miyagi 2016). On the other hand, mangrove distribution areas in Japan are in the path of typhoons, which cause damage such as fallen trees and broken trunks due to strong winds (Uchiyama and Miyagi 2020; Watanabe et al. 2022).

K. Fujimoto (✉)
Faculty of Policy Studies, Nanzan University, Nagoya, Japan
e-mail: kfuji@nanzan-u.ac.jp

M. Clüsener-Godt et al. (eds.), *Blue Carbon Mangrove Ecosystems*, Blue Carbon Ecosystems for Sustainable Development 1,
https://doi.org/10.1007/978-3-031-69553-7_11

The purpose of this paper is to introduce the carbon storage abilities of Japanese mangrove ecosystems, taking into account their geomorphic habitat conditions and the impact of typhoons, based on data obtained mainly from the subtropical Iriomote Island, southwestern Japan, comparing them with tropical mangrove ecosystems, especially Pohnpei Island, Federated States of Micronesia, where the author have studied on mangrove forest dynamics since 1994 with his colleagues (Fujimoto et al. 2023a).

11.2 Mangrove Distribution in Japan

The distribution of mangrove forests in Japan was first summarized by Nakasuga (1979). The latest information is available on the web site of Manglobal, which is a Japanese specified nonprofit organization established in 2003 (https://www.manglobal.or.jp/). The data on this website is based on steady field research and is highly reliable.

Figure 11.1 shows mangrove species found on the major islands in Nansei Islands of Japan located between Kyushu and Taiwan, and northern limit of mangrove distribution in the Pacific.

Kiire is famous place as the northern limit of mangrove distribution in the Pacific Ocean, but recently, it was confirmed to be distributed in the Kaminokawa (31.6191° N, 130.33243°E) by Manglobal (Fig. 11.1). Mangrove forest area in Japan is estimated to be only 885 ha, of which 604 ha is distributed on Iriomote Island (Manglobal: https://www.manglobal.or.jp/forest/).

Seven species of true mangroves are distributed in Japan, which are *Kandelia obovata*, *Bruguiera gymnorrhiza*, *Rhizophora stylosa*, *Avicennia marina*, *Sonneratia alba*, *Lumnitzera racemosa*, and *Nypa fruticans*. Only *K. obovata* is found at the northern limit of mangrove distribution. The northern limit of *B. gymnorrhiza* is Amami-Oshima Island, and the northern limit of *R. stylosa* and *A. marina* is Okinawa Island. All seven species are distributed on Iriomote Island, and six species except *N. fruticans* can be seen on Ishigaki Island (Fig. 11.1).

11.3 Aboveground Biomass of Japanese Mangrove Forests

Figure 11.2 shows the aboveground biomass of the main communities on Iriomote Island. IFR is a fixed plot 5 m wide and 45 m long installed in Funaura Bay in northern Iriomote Island in 2013 (except for 10 m seaward of IFR1 in Fujimoto et al. 2021), IY_{0-50} is a seaward 50 m section of a fixed plot 5 m wide and 70 m long installed in a mangrove forest facing Yubu Island in 2018 (plot code: IY), IY_{50-70} is a landward 50–70 m section of IY, and INB is a 5 m wide and 45 m long fixed plot installed in 2000 in a mangrove forest on the delta of the Nakama River in southeastern Iriomote Island. Aboveground biomass in Fig. 11.2 represents in the

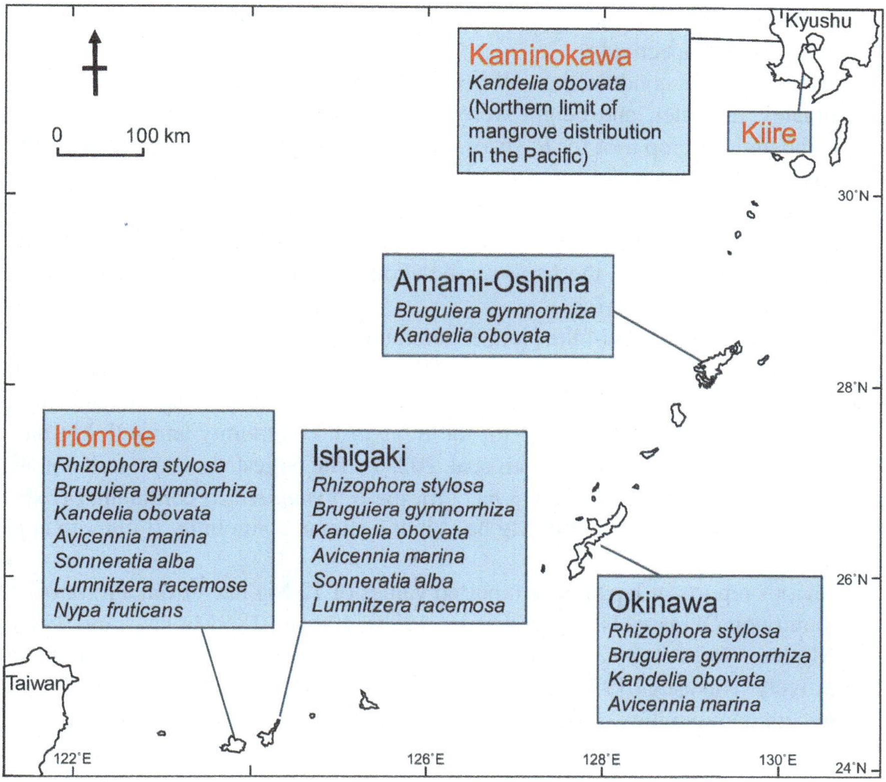

Fig. 11.1 Mangrove species found on the major islands in Nansei Islands of Japan and the northern limit of mangrove distribution in the Pacific

Fig. 11.2 Aboveground biomass of the major mangrove communities on Iriomote Island. See the text for the identification of plot codes on the horizontal axis

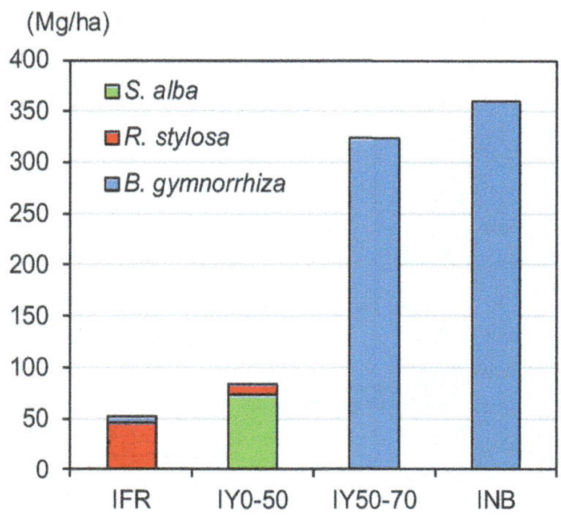

latest census results for each plot (IFR: July 2023, IY: December 2021, and INB: February 2018, respectively).

Tree census was conducted on all trees 1.3 m or taller, and tree species, location coordinates, diameter, and height were recorded. Diameter was measured 30 cm above the highest prop root for *R. stylosa* and breast height (1.3 m above the ground surface) for all other species.

Aboveground biomass was calculated using the common allometric equation by Komiyama et al. (2005) except for *R. stylosa*. Because this equation does not include the prop root biomass, so the part-specific allometric equation of Komiyama et al. (1988) was used for *R. stylosa*.

Aboveground biomass of the pioneer communities of *R. stylosa* (IFR) and *S. alba* ($IY_{0–50}$) was estimated to be 52 and 84 Mg ha^{-1}, respectively. These are about one-sixth of the values obtained from Pohnpei Island, Micronesia in the humid tropics, which were 327 Mg ha^{-1} for the *R. stylosa* community and 495 Mg ha^{-1} for the *S. alba* community (Fujimoto et al. 2023a). The largest aboveground biomass on Iriomote found so far is 361 Mg ha^{-1} for the *B. gymnorrhiza* community (INB), while on Pohnpei it reached 840 Mg ha^{-1} in the climax community (Fujimoto et al. 2023a).

Previous reports in Japan have reported values of 72 Mg ha^{-1} from a *K. obovata* community on Okinawa Island (Nakasuga 1979), 108 and 185 Mg ha^{-1} and 98 and 112 Mg ha^{-1} from *R. stylosa* and *B. gymnorrhiza* communities on Ishigaki Island, respectively (Nakasuga 1979; Suzuki and Tagawa 1983), and 32–81 Mg ha^{-1} from *K. obovata* communities on Amami Oshima Island (Ishihara et al. 2004).

11.4 Dynamics of Tree Density and Aboveground Biomass

Figure 11.3 shows the dynamics of tree density and aboveground biomass of *R. stylosa* community (IFR) and *B. gymnorrhiza* community (INB) on Iriomote Island.

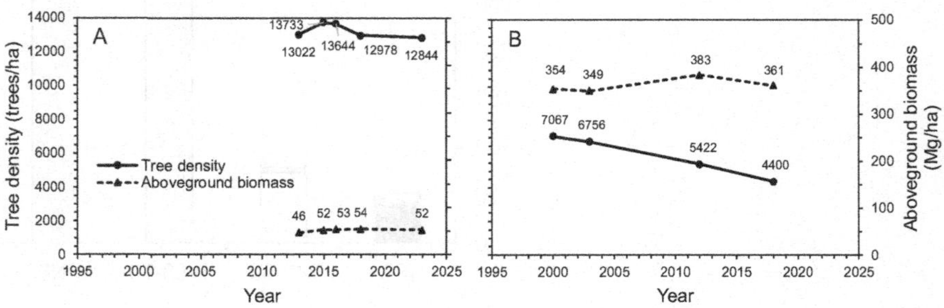

Fig. 11.3 Dynamics of tree density and aboveground biomass of (**A**) *Rhizophora stylosa* community (IFR) and (**B**) *Bruguiera gymnorrhiza* community (INB) on Iriomote Island

Tree density in IFR increased between 2013 and 2015, but has been declining since then (Fig. 11.3A). In particular, it decreased significantly between 2016 and 2018. On the other hand, tree density in the INB has consistently decreased since 2000 (Fig. 11.3B).

Aboveground biomass in IFR increased between 2013 and 2018, but declined between 2018 and 2023. The average accretion rate of aboveground biomass between 2013 and 2023 was calculated to be 0.6 Mg ha^{-1} $year^{-1}$, but for the period 2013–2015, it was calculated to be 3.1 Mg ha^{-1} $year^{-1}$. On the other hand, aboveground biomass in the INB increased between 2003 and 2012 despite a decrease in tree density, but decreased between 2012 and 2018. The average accretion rate of aboveground biomass between 2000 and 2018 was calculated to be 0.5 Mg ha^{-1} $year^{-1}$, but for the period 2003–2012, it was calculated to be 3.8 Mg ha^{-1} $year^{-1}$.

The marked decrease in tree density and aboveground biomass may be due to typhoons. Iriomote Island is regularly hit by typhoons with very strong forces. In particular, Typhoon No. 15 in August 2015 struck with a maximum wind velocity of over 54 m, causing significant tree damage to the Nakama River mangrove forest (Uchiyama and Miyagi 2020). The decrease in tree density and aboveground biomass between 2012 and 2018 in the INB is likely effect of the typhoon. Furthermore, typhoons with maximaum wind velocity of approximately 40 m also hit Iriomote Island in 2006, 2007, 2016 and 2018, and these typhoon strikes may also have temporarily reduced aboveground biomass. Compared with the average accretion rate of aboveground biomass of Pohnpei Island, 5.7 Mg ha^{-1} $year^{-1}$ for *R. stylosa* community and 6.8 Mg ha^{-1} $year^{-1}$ for the community with the largest aboveground biomass (Fujimoto et al. 2023a), the values for Iriomote are about one-tenth, but about one-half during periods of low typhoon impact.

11.5 Belowground Stored Carbon

Belowground stored carbon varies greatly depending on tree species (Fig. 11.4). In Southeast Asia, mangrove peat was found only in the *Rhizophora* communities (Fujimoto and Miyagi 1993; Mochida et al. 1999; Lo et al. 2018). Similarly, on Iriomote Island, mangrove peat can be found only in the *R. stylosa* communities, which were found on a relatively low elevation back marshs away from the river channel (Fujimoto and Ohnuki 1995; Fujimoto et al. 2002).

The mangrove peat layer accumulates 500 to 650 Mg ha^{-1} of carbon up to 1 m deep, and even on Iriomote in the subtropics, approximately 500 Mg C ha^{-1} is accumulated (Fig. 11.4). This means that even in the subtropics where tree size is smaller than in the tropics and productivity is lower, if the production rate exceeds the decomposition rate and enough time passes, then the systems have the ability to accumulate similar amounts of soil organic carbon as in the tropics (Fujimoto et al. 2023b).

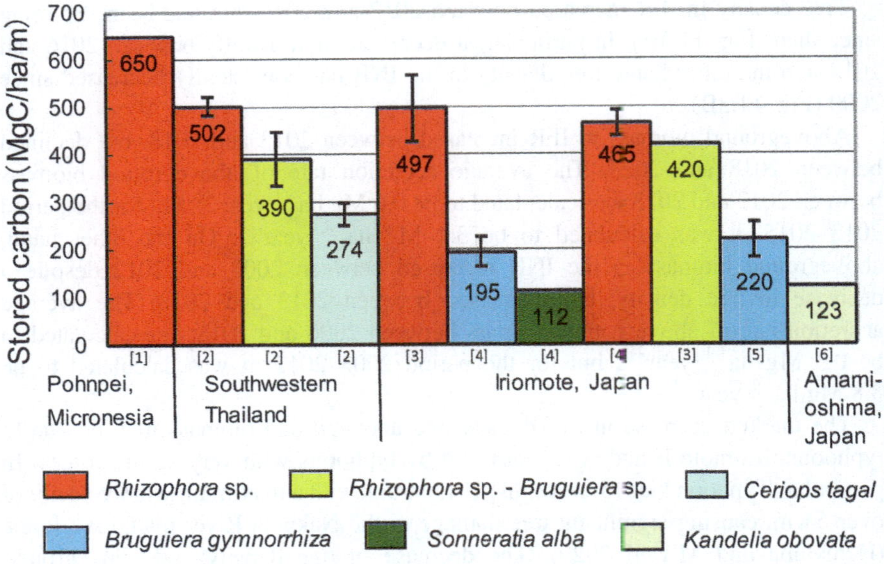

Fig. 11.4 Belowground stored carbon by community up to 1 m deep in the Asia-Pacific region (after Fujimoto et al. 2023b). Numbers in square brackets at the bottom of each columnar diagram indicate the literature of the data source; 1: Fujimoto et al. (1999), 2: Fujimoto et al. (2000a), 3: Fujimoto et al. (2000b), 4: Fujimoto et al. (2002), 5: Fujimoto et al. (1996), 6: Ishihara et al. (2004)

The *B. gymnorrhiza* community on Iriomote is distributed mainly on natural levees consisting of sandy deposits along rivers, and the *S. alba* community is also distributed on sandy deposits on the sandbars and seaward front of mangrove forests (Kikuchi et al. 1978; Fujimoto and Ohnuki 1995). The belowground stored carbon is approximately 200 Mg C ha^{-1} for *B. gymnorrhiza* communities and only 112 Mg C ha^{-1} for a *S. alba* community (Fujimoto et al. 2002). For the *K. obovata* community, 123 Mg C ha^{-1} was obtained from the Sumiyo River delta on Amami-Oshima (Ishihara et al. 2004).

Mangrove habitat on the delta is controlled by sediment in flow from rivers. This means that the sediment contains not only autochthonous carbon originating from within the mangrove ecosystem, but also a certain amount of allochthonous carbon imported from the catchment area. In addition, as the mangrove forests advance seaward with the development of the delta, the former mangrove deposits will be isolated inland. Furthermore, mangrove forests on Iriomote Island have shifted their habitat seaward due to seismic uplift caused by huge earthquakes during the late Holocene (Fujimoto and Ohnuki 1995; Fujimoto et al. 2015). This geomorphic process also leads to the isolation of former mangrove deposits on the inland side, though some organic matter near the surface may be decomposed after the uplift, and the accumulation of new mangrove deposits on the seaward side. In order to evaluate the belowground stored carbon of mangrove origin, such geomorphic processes must be taken into account.

11.6 Fine Root Production

The main parent material of mangrove peat is very fine roots (Ono et al. 2022). Therefore, we investigated fine root production by tree species using the ingrowth core method (Fujimoto et al. 2021). Fujimoto et al. (2021) developed a method to estimate the amount of decomposed fine roots during the incubation period of the ingrowth core by combining the litter bag method. Previous studies using the ingrowth core method ignored the missing dead roots (e.g., Castañeda-Moya et al. 2011; Adame et al. 2014; Cormier et al. 2015; Noguchi et al. 2020).

As a result, the annual fine-root production values up to a 50 cm depth were estimated at 16.91 Mg ha^{-1} $year^{-1}$ for a seashore *R. stylosa*, 16.58 Mg ha^{-1} $year^{-1}$ for an interior *R. stylosa*, 11.50 Mg ha^{-1} $year^{-1}$ for a seashore *B. gymnorrhiza* and 9.62 Mg ha^{-1} $year^{-1}$ for an interior *B. gymnorrhiza*, of which the decomposed roots were estimated at 1.82 Mg ha^{-1} $year^{-1}$, 2.39 Mg ha^{-1} $year^{-1}$, 1.89 Mg ha^{-1} $year^{-1}$ and 1.72 Mg ha^{-1} $year^{-1}$, respectively (Fujimoto et al. 2021).

When standardized to a layer thickness of 10 cm for comparison with the fine-root production of Pohnpei Island (Ono et al. 2022), which was estimated using the same method as Fujimoto et al. (2021), the fine-root production on Iriomote was clearly lower: about 25% of Pohnpei's on the seaward site of *R. stylosa*, about 52% on the landward site, and about 12% on the seaward site of *B. gymnorrhiza*, and 42% on the landward site (Fig. 11.5A).

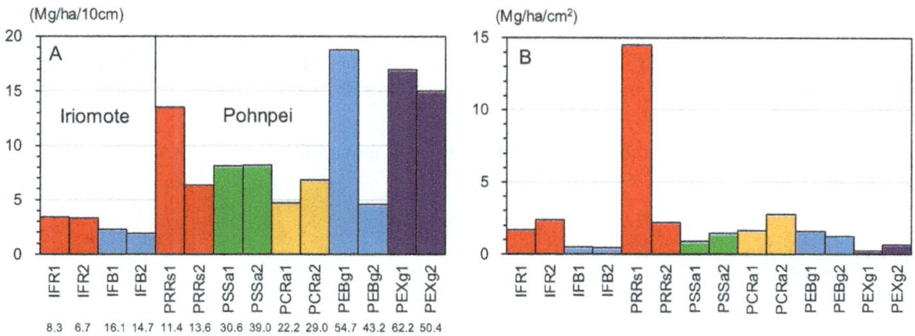

Fig. 11.5 (**A**) Annual fine-root production normalized to 10 cm thickness for major mangrove species of Iriomote Island, southwestern Japan and Pohnpei Island, Federated States of Micronesia and (**B**) Annual fine-root production normalized to 10 cm thickness relative to annual increase in stem basal area. [Plot code] IFR1: *Rhizophora stylosa* seaward site on Iriomote, IFR2: *R. stylosa* landward site on Iriomote, IFB1: *Bruguiera gymnorrhiza* seaward site on Iriomote, IFB2: *B. gymnorrhiza* landward site on Iriomote, PRRs1: *R. stylosa* seaward site on Pohnpei, PRRs2: *R. stylosa* landward site on Pohnpei, PSSa1: *Sonneratia alba* seaward site on Pohnpei, PSSa2: *S. alba* landward site on Pohnpei, PCRa1: *Rhizophora apiculata* seaward site on Pohnpei, PCRa2: *R. apiculata* landward site on Pohnpei, PEBg1: *B. gymnorrhiza* seaward site on Pohnpei, PEBg2: *B. gymnorrhiza* landward site on Pohnpei, PEXg1: *Xylocarpus granatum* seaward site on Pohnpei, PEXg2: *X. granatum* landward site on Pohnpei. The number under each plot code in (A) indicates the diameter (cm) of the sample tree. Created using data from Fujimoto et al. (2021) for Iriomote and Ono et al. (2022) for Pohnpei

However, the amount of fine-root production is greatly affected by tree size. Sample tree diameters vary widely as noted Fig. 11.5A. Therefore, to eliminate this effect, we compared the fine-root production relative to the increase in stem basal area. Then, it becomes clear that the ability of fine-root production of *Rhizophora* species is larger than that of other species (Fig. 11.5B). This trend is particularly apparent in the comparison between *R. stylosa* and *B. gymnorrhiza* on Iriomote, and between the seaward side of *R. stylosa* and other tree species on Pohnpei.

11.7 Conclusion

Japanese mangrove forests, located at the northern limit of mangrove distribution, have small tree size and low aboveground biomass, but belowground stored carbon, especially for *R. stylosa* communities with mangrove peat, is equivalent to that in the tropics. The factors contributing to the small aboveground biomass are thought to be the effects of frequent and intense typhoons in addition to temperature conditions. Although the rate of fine-root production, the parent material of mangrove peat, on Iriomote Island is 25–50% of that in the tropics, it can be evaluated that mangrove forests in Japan, especially *R. stylosa* forests, play a significant role as a place for carbon fixation as in the tropics.

Acknowledgements This paper is based on research conducted by the author in mangrove forests in Japan and Micronesia over a period of more than 30 years. The author would like to thank the many research colleagues who have worked with the author on these surveys. Recent study has been financially supported by JSPS Kakenhi Grant Numbers JP25282084, JP17H02034 and JP22H00757, and Nanzan University Pache Research Subsidy I-A-2 (FY 2023).

Conflicts of Interest The author declared that there are no conflicts of interest.

References

Adame MF, Teutli C, Santini NS et al (2014) Root biomass and production of mangroves surrounding a karstic oligotrophic coastal lagoon. Wetlands 34:479–488. https://link.springer.com/article/10.1007/s13157-014-0514-5

Castañeda-Moya E, Twilley RR, Rivera-Monroy VH et al (2011) Patterns of root dynamics in mangrove forests along environmental gradients in the Florida coastal everglades, USA. Ecosystems 14:1178–1195. https://link.springer.com/article/10.1007/s10021-011-9473-3

Cormier N, Twilley RR, Ewel KC et al (2015) Fine root productivity varies along nitrogen and phosphorus gradients in high-rainfall mangrove forests of Micronesia. Hydrobiologia 750:69–87. https://link.springer.com/article/10.1007/s10750-015-2178-4

Fujimoto K, Miyagi T (2016) Relationship between mangrove zonation and micro-landforms and its application. In: Fujimoto K et al (eds) Micro-landforms: the key to connect humans and nature. Kokon-Shoin Publisher, Tokyo, pp 80–104. (In Japanese)

Fujimoto K, Miyagi T (1993) Development process of tidal-flat type mangrove habitats and their zonation in the Pacific Ocean: a geomorphological study. Vegetatio 106:137–146. https://link.springer.com/article/10.1007/BF00045067

Fujimoto K, Ohnuki Y (1995) Developmental processes of mangrove habitat related to relative sea-level changes at the mouth of the Urauchi River, Iriomote Island. Southwestern Japan Q J Geogr 47:1–12. https://doi.org/10.5190/tga.47.1

Fujimoto K, Sakai H, Imaya A (1996) Evaluation of carbon accumulation function in sediments of mangrove forests: a case study in Iriomote Island, Japan. In: Proceedings of the general meeting of the Association of Japanese Geographers, pp 128–129. (In Japanese)

Fujimoto K, Imaya A, Tabuchi R et al (1999) Belowground carbon storage of Micronesian mangrove forests. Ecol Res 14:409–413. https://esj-journals.onlinelibrary.wiley.com/doi/10.1046/j.1440-1703.1999.00313.x

Fujimoto K, Miyagi T, Murofushi T et al (2000a) Evaluation of the belowground carbon sequestration of estuarine mangrove habitats, Southwestern Thailand. In: Miyagi T (ed) Organic material and sea-level change in mangrove habitat. Tohoku Gakuin University, Sendai, pp 101–109

Fujimoto K, Ishihara S, Yukino I et al. (2000b) Evaluation of carbon sequestration in mangrove ecosystem with its habitat dynamics in Iriomote Island, southwestern Japan. In: Proceedings of the general meeting of the Association of Japanese Geographers, p 60–61. (In Japanese)

Fujimoto K, Miyagi T, Kawaguchi S et al. (2002) Carbon sequestration in mangrove ecosystems with its habitat dynamics in Iriomote Island, Southwestern Japan. In: Research Institute for Subtropical Regions (ed) research report on mangroves. Naha, p 309–319. (In Japanese with English abstract)

Fujimoto K, Miura M, Haruyama S (2015) Mangrove habitat dynamics and seismic crustal movement during the late Holocene on the Nakama River Delta, Iriomote Island, southwestern Japan. Mangrove Sci 9:3–15. http://www.mangrove.or.jp/gakkai/subpage/gakkaishi/Mangrove_Science/Vol.9/Mangrove_Science_Vol.9;3-15.2015.pdf (In Japanese with English abstract)

Fujimoto K, Ono K, Watanabe S et al (2021) Estimation of probable annual fine root production and missing dead roots associated with the ingrowth core method: attempt with major mangrove species on Iriomote Island, southwestern Japan, situated in the subtropics. Mangrove Sci 12:11–24. http://www.mangrove.or.jp/gakkai/subpage/gakkaishi/Mangrove_Science/Vol.12/Mangrove_Science_Vol.12;11-24.2020.pdf

Fujimoto K, Ono K, Tabuchi R et al (2023a) Findings from long-term monitoring studies of Micronesian mangrove forests with special reference to carbon sequestration and sea-level rise. Ecol Res 38:494–507. https://doi.org/10.1111/1440-1703.12346

Fujimoto K, Furukawa K, Ono K et al (2023b) Effects of sea-level rise on blue carbon stocks of mangrove ecosystems: insights from Pohnpei Island. Federated States of Micronesia Carbon Footprint 2:15. https://www.oaepublish.com/cf/article/view/cf.2023.12

Ishihara S, Fujimoto K, Kawanishi M et al (2004) Relationship between mangrove vegetation and micro-topography and carbon storage of *Kandelia candel* (L.) Druce forests in Amani-Ohshima Island, Southwestern Japan. Jpn J For Environ 46:9–14. https://www.jstage.jst.go.jp/article/jjfe/46/1/46_KJ00005291914/_pdf/-char/ja (In Japanese with English abstract)

Kikuchi T, Tamura T, Makita H et al (1978) Arrangement of Swamp forests and landforms of the alluvial plain in the lower Nakama River, Iriomote Island, Ryukyu Archipelago, Japan: I. Mangrove Swamp. Ann Tohoku Geog Assoc 30:71–81. (In Japanese with English abstract)

Komiyama A, Moriya H, Prawiroatmodjo S et al (1988) Primary productivity of mangrove forest. In: Ogino K, Chihara M (eds). Biological system of mangrove: a report of East Indonesian mangrove expedition, 1986. Ehime University, pp 97–117.

Komiyama A, Poungparn S, Kato S (2005) Common allometric equations for estimating the tree weight of mangroves. J Trop Ecol 21:471–477. https://doi.org/10.1017/S0266467405002476

Lo J, Quoi LP, Visal S (2018) Some preliminary observations on peat-forming mangroves in Botum Sakor, Cambodia. Mires Peat 22:1–10. http://mires-and-peat.net/media/map22/map_22_03.pdf

Mochida Y, Fujimoto K, Miyagi T et al (1999) A phytosociological study of the mangrove vegetation in the Malay Peninsula: special reference to the micro-topography and mangrove deposit. Tropics 8:207–220. https://doi.org/10.3759/tropics.8.207

Nakasuga T (1979) Analysis of the mangrove stand. The science bulletin of the College of Agriculture, University of the Ryukyus 46:413–513. (In Japanese with English abstract)

Noguchi K, Poungparn S, Umnouysin S et al (2020) Biomass and production rates of fine roots in two mangrove stands in southern Thailand. JARQ 54:349–360. https://www.jircas.go.jp/sites/default/files/jarq/jarq/jarq54-4_349-360.pdf

Ono K, Fujimoto K, Hirata Y et al (2022) Total fine root production of several types of tropical mangrove forests evaluated by the continuous inflow estimation method in Pohnpei Island, Micronesia: large contribution of necromass production to the stocks of blue carbon originated from mangrove ecosystems. Ecol Res 37:33–52. https://doi.org/10.1111/1440-1703.12280

Suzuki E, Tagawa E (1983) Biomass of a mangrove forest and a sedge marsh on Ishigaki Island, South Japan. Jpn J Ecol 33:231–234. https://doi.org/10.18960/seitai.33.2_231

Uchiyama S, Miyagi T (2020) Mangrove forest dynamics of the Nakama River Delta, Iriomote Island, Japan: quantitative estimations based on the image archives during the recent 70 years. Mangrove Sci 11:3–15. http://www.mangrove.or.jp/gakkai/subpage/gakkaishi/Mangrove_Science/Vol.11/Mangrove_Science_Vol.11;3-15.2020.pdf (In Japanese with English abstract)

Watanabe S, Uchiyama S, Miyagi T (2022) Drone research revealed mangrove forest collapse damage on Iriomote Island. Mangrove Sci 13:3–8. http://www.mangrove.or.jp/gakkai/subpage/gakkaishi/Mangrove_Science/Vol.13/Mangrove_Science_Vol.13;3-8.2022.pdf (In Japanese with English abstract)

Chapter 12
Malaysia's Blue Carbon Overview

Waseem Razzaq Khan, Roland Kueh Jui Heng, Jane Koh, Malcom Demies, Abdul Razaque Memon, and Faridah Hanum Ibrahim

Abstract Climate change is one of the greatest challenges in the world and attracted many countries to mitigate greenhouse gas emissions through several mechanisms. Blue carbon has captured great interest recently. Malaysia has untapped blue carbon especially mangrove forests. The online review suggested that there are still a limited number of scientific studies on the overall aspect of blue carbon, though some reports on carbon storage in mangrove forests. Malaysia is highly committed to this matter through various international signatories and national level initiates such as through the establishment of the Malaysian Climate Action Council, Twelve

W. R. Khan (✉)
Department of Forestry Science, Faculty of Agricultural and Forestry Sciences, Universiti Putra Malaysia, Bintulu, Sarawak, Malaysia

Institut Ekosains Borneo, Universiti Putra Malaysia, Bintulu, Sarawak, Malaysia

Faculty of Forestry and Environment, Universiti Putra Malaysia, Serdang, Selangor Darul Ehsan, Malaysia
e-mail: khanwaseem@upm.edu.my

R. K. J. Heng (✉)
Department of Forestry Science, Faculty of Agricultural and Forestry Sciences, Universiti Putra Malaysia, Bintulu, Sarawak, Malaysia
e-mail: roland@upm.edu.my

J. Koh
Faculty of Forestry and Environment, Universiti Putra Malaysia, Serdang, Selangor Darul Ehsan, Malaysia

Southridge Malaysia Sdn Bhd, Kuala Lumpur, Malaysia

M. Demies
Forest Department Sarawak Headquarters, Kuching, Sarawak, Malaysia

A. R. Memon
Huawei Tech United Arab Emirates, FZ LLC, Dubai, United Arab Emirates

F. H. Ibrahim
Institute of Tropical Forestry and Forest Products (INTROP), Universiti Putra Malaysia, Serdang, Selangor Darul Ehsan, Malaysia

M. Clüsener-Godt et al. (eds.), *Blue Carbon Mangrove Ecosystems*, Blue Carbon Ecosystems for Sustainable Development 1,
https://doi.org/10.1007/978-3-031-69553-7_12

Malaysian Plan, and National Policy on Biodiversity. Despite these, Malaysia still requires extensive data and research to realize the blue carbon potential. Strong coordination and cooperation between Federal/State, amendment to current law, and incorporation of GIS and Lidar technology in research. With these improvements, it is hoped that Malaysia can capitalize on the blue carbon potential and enhance its preservation and protection.

Keywords Blue carbon · Mangrove forest · Seagrass · Salt marsh · Malaysia

12.1 Introduction

Today, climate change is one of the greatest challenges in the world. Climate change occurring across the globe has sparked serious interest in mitigating greenhouse gas emissions through several mechanisms such as international commitment (United Nations Framework Convention on Climate Change or UNFCCC) and participatory programs such as carbon reduction programs through restoration/reforestation, Reduced Emissions from Deforestation and Forest Degradation (REDD+) and Clean Development Mechanism (CDM). Many countries have pledged to commit to reducing their countries' emissions. Malaysia is set to cut carbon intensity against GDP by 45% by 2030 compared to 2005 levels. After these commitments, countries will need to find ways to reduce their dependency on fossil fuels. As suggested by many, one of the solutions is carbon markets. This provides a platform for businesses and individuals to buy and sell carbon credits.

Recently, scientists have termed a full spectrum of colours for carbon namely brown, black, blue, red, teal, and green (Zinke 2020). Brown carbon is emission from the burning of fossil fuels and releasing CO_2 while black carbon arises from incomplete combustion of fossil fuels, biofuel, and biomass. Both carbons are major contributors to global warming. In contrast, green carbon is carbon removed by photosynthesis and stored in terrestrial plants and soil of the natural ecosystem, while blue carbon is the carbon captured by the oceans (Nellemann et al. 2009; IPCC 2021). Red carbon is the newest colour which refers to all biological particles on snow and ice that reduce albedo to survive, while teal carbon relates to carbon stored in inland freshwater wetlands (Zinke 2020).

Natural fixes using the natural ecosystem to capture and store carbon are among the suggested solutions by many scientists. Trumper et al. (2009) reported that if the forest, peatlands, and agriculture ecosystems are safeguarded and their' carbon restored, it may help to reduce over 50 gigatonnes of carbon emissions that would otherwise enter the atmosphere. This also includes grasslands and coastal ecosystems like mangroves (Khan and Aljahdali 2022). This is because it is estimated about 20% of greenhouse gas emissions are sourced from the clearing and burning of forests.

While many are concerned with the declining green carbon ecosystem as well as there is a need also to promote interest in the blue carbon ecosystem. This is because blue carbon has been reported to have captured more than 55% of green carbon.

These are stored in the form of (i) sediments from mangroves, (ii) salt marshes, and (iii) seagrasses (Nellemann et al. 2009). Despite this, some scientists classified blue carbon into (i) deep blue carbon which is in the high seas, where carbon is stored in continental shelf waters, deep sea waters, and sea floor, and (ii) coastal blue carbon focuses on the coastal zone (mangroves, seagrass, and tidal marshes) (UNEP 2017; Twilley and Rovai 2019; Ocean Frontier Institute 2023; Rastogi and Siegel 2023).

The decline and degradation of coastal and marine ecosystems can release the captured carbon back into the atmosphere, which contributes to climate change and reverses the climate adaptation mitigation benefits they provide. According to Rastogi and Siegel (2023), global emission reduction alone is insufficient to achieve net zero but needs to include carbon dioxide removal (CDR) effort. CDR has been focused on terrestrial-based such as reforestation, soil capture, and technology-based direct-air capture. The opportunity of ocean-based CDR or marine CDR (mCDR) should not be dismissed as the ocean is the largest carbon storage sink on earth.

In the Malaysian context, mangroves are considered the front runner for blue carbon research due to their difficulty in reaching salt marshes and seagrasses meadows. A report by Omar et al. (2020) suggested that the largest mangrove areas are found in Sabah (60%), followed by Sarawak (22%) and 18% in Peninsular Malaysia with a total area estimated at 629,038 ha in 2017. The estimated average of Malaysia's mangrove carbon stock (above- and below ground) is about 181 MgC/ha. Between 1990 to 2017, the loss of mangrove area accounted for 3.3% which contributes to about 14 million Mg CO_2 emissions. Although the report also suggested caution that the figures obtained are generally crude, it provides an idea for further studies to about detailed studies related to carbon cycle/storage in the mangrove forest.

12.2 Research Studies in Malaysia

A review by Taillardat et al. (2018) found that global blue carbon ecosystems have the highest carbon sinks per plot compared to other types of ecosystems. It is suggested that the highest carbon sink is in salt marshes, followed by mangroves and seagrasses. Despite this, it is also highlighted the lack of data limits the assessment of the role of blue carbon. Therefore, articulating the role of this blue carbon in climate change mitigation, it is essential to address the data limitation.

In Malaysia, such a scenario is similar where some of the accessible literature via Google Scholar search was summarized in Table 12.1. Only literature related to the above, below ground, soil biomass/carbon of mangrove forests while other components of the carbon pool in this blue carbon context are relatively limited. The salt marsh is a coastal ecosystem in the upper coastal intertidal zone between land and open saltwater and is dominated by salt-tolerant plants. Seagrass is a flowering plant rooted in sediment on the sea bottom with shoots appearing above the substrate which are an important community in the shallow marine environment together with seaweeds and phytoplankton (Japar Sidik et al. 2012). Japar Sidik et al. (2012) made

Table 12.1 Information on the carbon storage at different blue carbon forests in Malaysia

	Total	Biomass (Mg/ha)	Carbon (MgC/ha)	Reference
1	Malaysia		181	Omar et al. (2020)
2	Peninsular Malaysia		151.40–895.8	Omar et al. (2016), Hong et al. (2017), Adame et al. (2018), and Rozainah et al. (2018a, b)
	Above ground			
1	Peninsular Malaysia	2.6–311.6	35.4–121.8	Tanouchi et al. (2000), Kasawani et al. (2007), Norhayati et al. (2009), Faridah-Hanum et al. (2012), Tengku Zarawie et al. (2015), Hong et al. (2017), Otero et al. (2018), Rozainah et al. (2018a, b), and Muhammad-Nor et al. (2019)
2	*Rhizophora apiculata*	460.0		Putz and Chan (1986)
3	*Bruguiera parviflora*	16.5		Hossain et al. (2007)
4	Mature *Rhizophora*		64.1	Hamdan et al. (2013)
	Logged-over *Rhizophora* (≤20 years)		49.5	Hamdan et al. (2013)
	Avicennia-Sonneratia mixed forest		18.2	Hamdan et al. (2013)
	30-year-old R. apiculata	372.0		Goessens et al. (2014)
	30-year-old R. apiculata	334.0		Hazandy et al. (2015)
	15-year-old R. apiculata	235.0		Khan et al. (2019)
	25-year-old R. apiculata	241.0		Khan et al. (2019)
	30-year-old R. apiculata	266.0		Khan et al. (2019)
5	Sabah	196.9	136.6	Wong et al. (2020) and Mhd Hatta et al. (2022)
	Sarawak			
6	*Rhizophora apiculata*	116.8		Chandra et al. (2011).
	Below ground			
1	Peninsular Malaysia	20.8–62.6	3.3–26.1	Hong et al. (2017), Rozainah et al. (2018a) and Muhammad-Nor et al. (2019)
2	Sabah		45.5	Mhd Hatta et al. (2022)
	Soil			
	Peninsular Malaysia			
1	0–100 cm depth		384.6–413.3	Rozainah et al. (2018a, b)

(continued)

Table 12.1 (continued)

	Total	Biomass (Mg/ha)	Carbon (MgC/ha)	Reference
2	0–20 cm depth		90.1–119.7	Hong et al. (2017)
	Sabah			
3	0–100 cm depth		273.8	Mhd Hatta et al. (2022)
	Seagrass			
1	Malaysia		108.6	Stankovic et al. (2021)

a review paper on seagrass and indicated that the subject matter is the least studied in Malaysia. They cited a few papers such as those by Ethirmannasingam et al. (1996), and Japar Sidik et al. (1996).

12.3 Malaysia's Commitment

Malaysia is highly committed to the issues related to climate change and carbon where engagement with the international community includes the signing of the UNFCCC in 1993 and Kyoto Protocol in 1999 and the Paris Agreement in 2016. Malaysia declared its Nationally Determined Contribution (NDC) under the Paris Agreement to reduce its economy-wide GDP-linked greenhouse gas (GHG) emission intensity by 45% in 2030 compared to the 2005 level. The first NDC (2015) included a condition that 10% of the emission intensity reduction is only committed upon international assistance. The updated NDC (2021) committed to a 45% emission intensity reduction unconditionally. They elucidate the highest level of government's commitment to the matter and tall order for the nation. The National Steering Committee on Climate Change (NSCC) was established in 1994 and the Malaysian Climate Change Action Council (MyCAC) was established in 2021 to formulate and implement policies to address and adapt to climate change.

Under the Twelfth Malaysia Plan (2021–2025), advancing sustainability was highlighted which is to guarantee continuous economic growth that balances the quality of life while at the same time protecting the environment, and conserving natural resources. Among others the key performance indicators used are (i) a 45% reduction in greenhouse gas (GHG) emissions intensity to GDP by 2030, (ii) 25% in government green procurement, (iii) at least 20% terrestrial and inland water areas conserved, (iv) at least 10% in coastal and marine area conserved.

These plans are in line with the National Policy on Biological Diversity 2016–2025 which policy envisions that by 2025 all production **forests, agriculture production, and fisheries will be managed and harvested sustainably. For example, by 2025, 100% of timber and products will be certified under the Malaysian Timber Certification Scheme (MTCS) scheme, while for agricultural areas under Malaysian Sustainable Palm Oil (MSPO), Roundtable on**

Sustainable Palm Oil (RSPO) and fisheries under Good Agricultural Practices (GAP), Marine Stewardship Council (MSC), etc.

Under this policy, it also hoped that by 2025 (i) at least 20% of terrestrial areas and inland waters, and 10% of coastal and marine areas, will be conserved through a representative system of protected areas and other effective area-based conservation measures, (ii) vulnerable ecosystems and habitats, particularly limestone hills, wetlands, coral reefs, and seagrass beds, are adequately protected and restored.

12.4 Challenges and Opportunities

In adapting the blue carbon concept and as mitigating measures towards climate change, Malaysia still requires extensive data and research to realize especially the actual coastal blue carbon potential. This will help in preparing Malaysia to be ahead of time to be one of the key players in the blue carbon system. This has been highlighted in a paper presented by Teh (2014). The mangrove ecosystem is also under threat posed by development projects and encroachment.

A national data repository is also an important tool for data storage and sharing which can evaluate the value and role of mangrove ecosystems as carbon sinks and CDR while national level toolkits or guidelines facilitate and standardize the research and reporting protocol. This provides coordinated efforts in the research and reporting of data suitable for the market. Eventually, such a data pool can assist in policy improvement and efforts toward quantifying the role of the blue carbon ecosystem in mitigating climate change.

Malaysian federal organizations and state governments are in the process establish establishing carbon laws. Prime Minister Datuk Seri Anwar Ibrahim has announced a ten million ringgit for a voluntary carbon market to show government seriousness to support and achieve net zero and with it, Bursa Malaysia has announced its Voluntary carbon market (VCM) to trade carbon credits. Further, Sarawak state amended its Forest Ordinance in 2022 and added important sections to support carbon trading (Ministry of Finance 2023).

With all these efforts, there are fewer attempts have been seen to establish laws and regulations for blue carbon. The Malaysian government has created a section to support blue carbon research but until now, no methodologies have been recommended by the government organizations to do further research on carbon quantification in blue carbon. So, by looking at this scenario, a research-based unit that will solely work on blue carbon needs time to support this ecosystem.

12.5 Conclusion

Malaysia has great potential for blue carbon research. Government attention is greatly needed to propel and coordinate efforts as there are many universities, NGOs, federal/state agencies, and federal/state government involved. Without these systematic and coordinated efforts, Malaysia will not be able to capitalize on the mangrove blue carbon forest in the country. Furthermore, GIS and Lidar tools can be useful to increase the data depository of blue carbon in this country. Specifically, coordination at the Federal and State levels in establishing a framework for blue carbon initiatives and methodologies is crucial in providing the foundation for future involvement in carbon trading and any other CDR programme. Amendment in the forestry and fisheries legislation at federal and state levels to legitimate carbon as produce so that it can be monetized. With these improvements, it is hoped that the blue carbon can be further preserved and protected under the national legal system.

Acknowledgement We acknowledge the support provided by the management of University Putra Malaysia and Forest Department Sarawak.

Conflict of Interest The authors declare no conflict of interest.

References

Adame MF, Zakaria RM, Fry B, Chong VC, Then YHA, Brown CJ, Lee SY (2018) Loss and recovery of carbon and nitrogen after mangrove clearing. Ocean Coast Manag 161:117–126

Chandra IA, Seca G, Abu Hena MK (2011) Aboveground biomass production of *Rhizophora apiculata* Blume in Sarawak Mangrove Forest. Am J Agric Biol Sci 6(4):469–474

Ethirmannasingam S, Phang SM, Sasekumar A (1996) A study on some phenological events in a Malaysian Enhalus acoroides beds. In: Kuo J, Phillips RC, Walker DI, Kirkman H (eds) Seagrass Biology: Proceedings of an International Workshop, Rottnest Island. Faculty of Science, The University of Western Australia, Nedlands, pp 33–40

Faridah-Hanum I, Abd Kudus K, Saari MS (2012) Plant diversity and biomass of Marudu Bay mangroves in Malaysia. Pak J Bot 44:151–156

Goessens A, Satyanarayana B, van der Stocken T, Zuniga MQ, Mohd-Lokman H, Sulong I, Dahdoub-Guebes F (2014) Is Matang mangrove forest in Malaysia sustainably rejuvenating after more than a century of conservation and harvesting management? PLoS One 9(8):e105069

Hamdan O, Khairunnisa MR, Ammar AZ, Mohd Hasmadi I, Aziz KH (2013) Mangrove carbon stock assessment by optical satellite imagery. J Trop For Sci 25(4):554–565

Hazandy AH, Ahmad AN, Zaiton S, Tuan M, Mohammad LS (2015) Quantification and economic valuation of carbon stock in the Matang Mangrove Forest Reserve in Perak. In: Proceedings of the 17th Malaysian forestry conference, a century of forest management: Lessons learnt & the way forward, Kota Kinabalu, Sabah, Malaysia, 11th–12th November, 2014. Sabah Forestry Department, Sabah, pp 321–335

Hong LC, Hemati Z, Zakaria R (2017) Carbon stock evaluation of selected mangrove forests in peninsular Malaysia and its potential market value. J Environ Sci Manag 20(2):77–87

Hossain M, Othman S, Bujang JS, Kusnan M (2007) Net primary productivity of *Bruguiera parviflora* (Wight & Arn.) dominated mangrove forest at Kuala Selangor, Malaysia. For Ecol Manag 255(1):179–182

IPCC (2021) Annex VII: Glossary. In: Masson-Delmotte V, Zhai P, Pirani A, Connors SL, Péan C, Berger S, Caud N, Chen Y, Goldfarb L, Gomis MI, Huang M, Leitzell K, Lonnoy E, Matthews JBR, Maycock TK, Waterfield T, Yelekçi O, Yu R, Zhou B (eds) Climate change 2021: the physical science basis. Contribution of working group I to the sixth assessment report of the intergovernmental panel on climate change. Cambridge University Press, Cambridge, pp 2215–2256

Japar Sidik B, Arshad A, Hishamuddin O, Muta Harah Z, Misni S (1996) Seagrass and macroalgal communities of Sungai Pulai estuary, South-East Johore, Peninsular Malaysia. In: Kuo J, Phillips RC, Walker DI, Kirkman H (eds) Seagrass Biology: Proceedings of an International Workshop, Rottnest Island. Faculty of Science, The University of Western Australia, Nedlands, pp 33–40

Japar Sidik B, Muta Harah Z, Kawaguchi S (2012) Historical review of seagrass research in Malaysia before 2001. Coastal Mar Sci 35(1):157–168

Kasawani I, Kamaruzaman J, Nurun-Nadhirah MI (2007) Biological diversity assessment of Tok Bali Mangrove Forest, Kelantan, Malaysia. WSEAS Trans Environ Dev 2(3):37–44

Khan WR, Aljahdali MO (2022) Elemental composition of above and belowground mangrove tissue and sediment in managed and unmanaged compartments of the Matang Mangrove Forest Reserve. Plan Theory 11(21):2916

Khan WR, Zulkifli SZ, Kasim MRB, Pazi AM, Mostapa R, Nazre M (2019) Mangrove productivity estimation using modelling approach and tree parameters assessment. Trop Conserv Sci 12: 1940082919872137

Mhd Hatta S, Salleh E, Suhaili NS, Besar NA (2022) Estimation of carbon pool at mangrove forest of Kudat, Sabah, Malaysia. Biodiversitas 23(9):4601–4608

Ministry of Finance (2023) RM10 mlm seed fund for Malaysian-generated carbon credit to kickstart market-PM Anwar. Retrieved on 26 July 2023 at https://www.mof.gov.my/portal/en/news/press-citations/rm10-mln-seed-fund-for-malaysian-generated-carbon-credit-to-kickstart-market-pm-anwar

Muhammad-Nor SM, Huxham M, Salmon Y, Duddy SJ, Mazars-Simon A, Mencuccini M, Meir P, Jackson G (2019) Exceptionally high mangrove root production rates in the Kelantan Delta, Malaysia: an experimental and comparative study. For Ecol Manag 444:214–224

Nellemann C, Corcoran E, Duarte CM, Valdes L, De Young C, Fonseca L Grimsditch G (2009) Blue carbon. A rapid response assessment. United Nations Environment Programme, GRID-Arendal. Birkeland Trykkeri

Norhayati A, Shukor MN, Juliana S, Wan Juliana WA (2009) Mangrove flora and fauna of Klang Islands Mangrove Forest Reserves, Selangor, Malaysia. Malays J Sci 28(3):275–288

Ocean Frontier Institute (2023) Deep blue carbon. Retrieved on 13 April 2023 at https://www.ofi.ca/impact/policy/deep-blue-carbon

Omar H, Mohd Johan Chuah N, Parlan I, Musa S (2016) Assessing rate of deforestation and changes of carbon stock on mangroves in Pahang, Malaysia. Malaysian For 79(1 & 2):174–179

Omar H, Husin TM, Parlan I (2020) Status of mangroves in Malaysia. Forest Research Institute Malaysia, Kepong

Otero V, Van de Kerchove R, Satyanarayana B, Martinez-Espinosa C, Fisol MA, Ibrahim MR, Sulong I, Mohd-Lokman H, Lucas R, Dahdouh-Guebas F (2018) Managing mangrove forests from the sky: forest inventory using field data and unmanned aerial vehicle (UAV) imagery in Matang mangrove forest reserve, peninsular Malaysia. For Ecol Manag 411:35–45

Putz FE, Chan HT (1986) Tree growth, dynamics, and productivity in a mature mangrove forest in Malaysia. For Ecol Manag 17(2):211–230

Rastogi A, Siegel E (2023) Advancing responsible deep blue carbon: a business strategist perspective. Retrieved on 13 April 2023 at https://impact.economist.com/ocean/ocean-and-climate/advancing-responsible-deep-blue-carbon-a-business-strategist-perspective

Rozainah MZ, Narzi MN, Sofawi AB, Hemato Z, Juliana WA (2018a) Estimation of carbon pool in soil, above and below ground vegetation at different types of mangrove forests in peninsular Malaysia. Mar Pollut Bull 137:237–245

Rozainah MZ, Sofawi AB, Joharee NA, Pauzi AZ (2018b) Stand structure and biomass estimation in the Klang Islands mangrove Forest, Peninsular Malaysia. Environ Earth Sci 77(13):486

Stankovic M, Ambo-Rappe R, Carly F, Dangan-Galon F, Fortes MD, Hossain MS, Kiswara W, Luong CV, Minh-Thu P, Mishra AK, Noiraksar T, Nurdin N, Panyawai J, Rattanachot E, Rozaimi M, Htun US, Prathep A (2021) Quantification of blue carbon in seagrass ecosystems of Southeast Asia and their potential for climate change mitigation. Sci Total Environ 783:146858

Taillardat P, Friess DA, Lupascu M (2018) Mangrove blue carbon strategies for climate change mitigation are most effective at the national scale. Biol Lett 14(10):20180251

Tanouchi H, Nakamura S, Ochiai Y, Azman H (2000) Profiles of species composition and aboveground biomass in a mangrove forest, Matang, Peninsular Malaysia. Japan Agr Res Q 34(2):147–151

Teh E (2014) Coastal blue carbon: exploring the opportunities and challenges for Malaysia. Poster presented at the International Conference on Mangroves of Asia-Pacific countries in view of climate change, 11–13 November, 2014. Grand Seasons Hotel, Kuala Lumpur

Tengku Zarawie TH, Suratman MN, Jaafar J, Mohd Hasmadi I, Abu F (2015) Field assessment of above ground biomass (AGB) of mangrove stand in Merbok, Malaysia. Malays Appl Biol 44(3):81–86

Trumper K, Bertzky M, Dickson B, van der Heijden G, Jenkins M, Manning P (2009) The natural fix? The role of ecosystems in climate mitigation. UNEP-WCMC, Cambridge

Twilley R, Rovai A (2019) Why protecting blue carbon storage is crucial to fighting climate change. Retrieved on 13 April 2023 at https://www.greenbiz.com/article/why-protecting-blue-carbon-storage-crucial-fighting-climate-change

United National Environment Programme (UNEP) (2017) Blue forests: finding coastal and marine solutions to meet the Paris Agreement. Retrieved on 13 April 2023 at https://www.unep.org/news-and-stories/story/blue-forests-finding-coastal-and-marine-solutions-meet-paris-agreement

Wong CJ, James D, Besar NA, Kamlun KU, Tangah J, Tsuyuki S, Phua M-H (2020) Estimating mangrove above-ground biomass loss due to deforestation in Malaysian Northern Borneo between 2000 and 2015 using SRTM and Landsat images. Forests 11(9):1018

Zinke L (2020) The colours of carbon. Nat Rev Earth Environ 1(3):141

Chapter 13
Blue Carbon Ecosystems, the 30 by 30 Challenge and the MAB Programme of UNESCO

Miguel Clüsener-Godt

Abstract The *30 by 30 Challenge* was launched in 2019 as a worldwide initiative for governments to designate 30% of Earth's land and ocean areas as protected areas by 2030, 10% as strictly protected areas. This *Challenge* aims to halt the accelerating loss of biodiversity and protect vital ecosystems that are the source of our economic security.

The 15th Conference of the Parties to the UN Convention on Biological Diversity, convened under the auspices of the United Nations, chaired by China, and hosted by Canada, adopted the "Kunming-Montreal Global Biodiversity Framework" (GBF), including four goals and 23 targets for achievement by 2030. This Framework was signed by 196 Government Parties.

Its main outcome consists of an agreement to the effective conservation and management of at least 30% of the world's land, coastal areas and oceans (currently, 17% of land and 8% of marine areas are under protection). It also foresees the restoration of 30% of terrestrial and marine ecosystems, the reduction to near zero of the loss of areas of high biodiversity importance and high ecological integrity.

UNESCO, with its network of around 2000 sites covering almost 6% of terrestrial surface, can play an important role in achieving this challenge in stimulating all its 194 Member States to adhere to the Kunming-Montreal Global Biodiversity Framework, and, at the same time, in nominating additional sites to UNESCO, so as to achieve 50% of the *Challenge*.

Keywords Blue carbon ecosystems · UNESCO · Man and the biosphere Programme (MAB) · Biosphere reserves · 30 by 30 challenge · Integrated nature conservation and sustainable development · Kunming-Montreal global biodiversity framework (GBF)

M. Clüsener-Godt (✉)
UNESCO Chair on Biodiversity Safeguard for Sustainable Development, CFE, Centre for Functional Ecology – Science for People & the Planet, Department of Life Sciences, University of Coimbra, Coimbra, Portugal
e-mail: clusenergodt@uc.pt

M. Clüsener-Godt et al. (eds.), *Blue Carbon Mangrove Ecosystems*, Blue Carbon Ecosystems for Sustainable Development 1,
https://doi.org/10.1007/978-3-031-69553-7_13

13.1 Blue Carbon Ecosystems

What do we know about carbon? There are four designations for carbon: Black, brown, green and blue carbon. Carbon occurs on Earth as black carbon mainly in fossil fuels and coals. Brown carbon is known for biomass but also for fossil fuel emissions. Green carbon designates all plant and chlorophyllin aspects of life. Finally, blue carbon is understood as coastal biomass and subhydric and semi-subhydric, such as mangroves, seagrasses, saltmarshes. However, today blue carbon in freshwater ecosystems such as peats and bogs add to the general definition of blue carbon.

13.2 Does the Amazon Produce the Oxygen That We Breathe?

In the entire discussion about the rise of CO_2 in the air and the noticed consequent rise of global temperature, there are many wrong messages circulating that lead to confusion. For example, that the Amazonian rainforest produces 20% of the world's oxygen. This message has frequently been circulated in the television news and in social media posts by politicians and celebrities, such as, for example, French president Emmanuel Macron, U.S. Vice President Kamala Harris, and the actor and environmentalist Leonardo di Caprio. However, several scientists have pointed out, the Amazon's net production to the oxygen that exists in the atmosphere is around zero. Michael Coe, who directs the Amazon program at the Woods Hole Research Center in Massachusetts, states clearly that "*There are a number of reasons why you would want to keep the Amazon in place, oxygen just isn't any one of them,*" and "*just doesn't make any physical sense*" because there is not enough CO_2 in the atmosphere to have synthesized 20% of the oxygen that exists in the atmosphere. The origin of the oxygen lies in the phytoplankton that has been accumulated in the sediments over billions of years and thus liberated the oxygen that we breathe (Zimmer, 2019).

13.3 Storage of Carbon in Ecosystems Today

According to IPCC and NASA, see Fig. 13.1, carbon is stored in the soil twice as much as in the atmosphere and in all living organisms combined. Carbon is well stored in the soil when its deterioration and consequent release in the atmosphere as CO_2 is stopped. Therefore, permafrost soils, as long as they do not melt, have a huge capacity of carbon storage, but the highest carbon storage capacity lies in wetlands due to their anaerobic conditions (see Fig. 13.1).

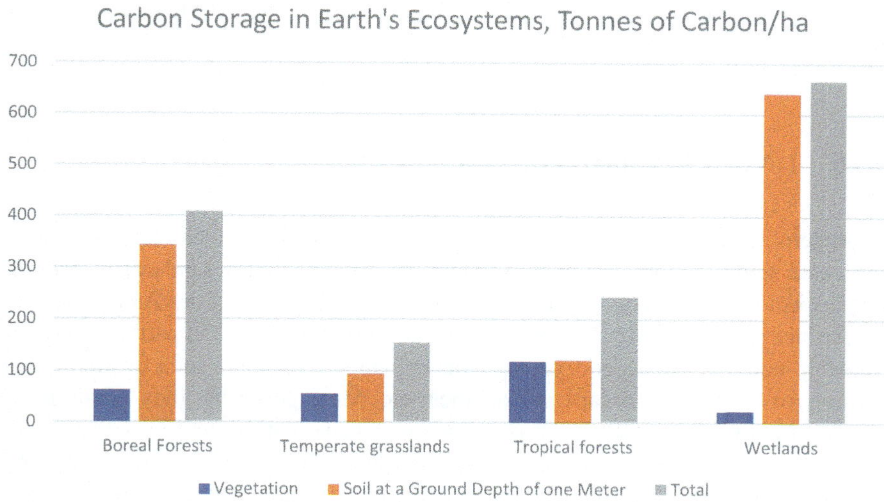

Fig. 13.1 Carbon storage in some of the earth's ecosystems (Visual Capitalist, January 2022). Visualizing carbon storage in earth's ecosystems, 2022, Dorothy Neufeld, graphics & design and Miranda Smith. Data source IPCC, NASA"

13.4 Risks for the Carbon Storage

Mainly due to land use changes, the carbon storage of ecosystems has become less efficient. This is a direct result of the overuse of the world's resources due to exponential increase of the population mainly since 1950, which more than doubled during this time span, passing the number of 8 billion in November 2022 and is predicted to arrive at 11 billion by the end of the century.

13.5 Ways Ahead: Creation and Sustainable Management of Protected Areas and the 30 by 30 Challenge

By establishing protected areas, the world's natural ecosystems can be substantially improved including the carbon storage capacities. However, main conservation agencies point out that only 20% of protected areas are being managed well (UNEP-WCMC, IUCN and NGS (2018). Protected Planet Report 2018). Protected areas must include land and sea, must always include in their management structures indigenous people and local communities, and dispose of adequate funding for their effective management. Although it is always less costly to protect than to restore nature areas, the degradation of any natural space should be avoided, not only in the protected areas but also in the areas that lie outside. Therefore, the establishment of integrated protected and sustainably managed areas is absolutely necessary, such as the UNESCO Biosphere Reserves, particularly in Blue Carbon Ecosystems (Clüsener-Godt et al. 2022).

13.6 UNESCO Biosphere Reserves

Biosphere Reserves have a three-fold zonation. Core Areas are legally strictly protected zones that are designed to ensure the conservation of landscapes, ecosystems, species, and genetic variation. Buffer Zones surround all Core Areas to buffer and reduce impacts. They also can serve conservation purposes, in particular if their valuable ecosystems depend on specific human intervention such as traditional farming or herding, landscape management etc. More generally, they are used for activities compatible with sound ecological practices that mutually reinforce with scientific research, monitoring, training, and education. Transition Areas are those parts of a Biosphere Reserve, which should not be governed by additional regulation but through incentives and innovation, such that communities are empowered to conduct socio-culturally and ecologically sustainable activities, in particular truly Green Economies. Biosphere Reserves include coasts and islands with surrounding marine areas (see Fig. 13.2).

The UNESCO World Network of Biosphere Reserves has 748 sites in 134 countries, including 23 transboundary biosphere reserves. The total size of all biosphere

Fig. 13.2 The zonation of a biosphere reserve. (UNESCO 2024a)

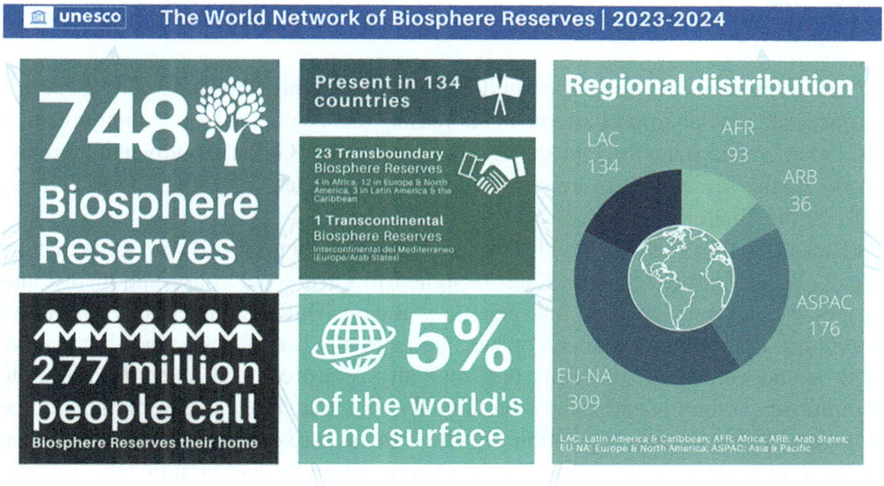

Fig. 13.3 The world network of biosphere reserves 2023–2024. (UNESCO 2024b)

reserves is more than seven million km^2, and 277 million people live in these areas (Fig. 13.3).

13.7 The 30 by 30 Challenge

The *30 by 30 Challenge* was launched in 2019 as a worldwide initiative for governments to designate 30% of Earth's land and ocean areas as protected areas by 2030, 10% as strictly protected areas. This *Challenge* targets to stop the loss of species and to protect all ecosystems.

The 15th Conference of Parties to the UN Convention on Biological Diversity, convened under the auspices of the United Nations, chaired by China, and hosted by Canada, adopted the "Kunming-Montreal Global Biodiversity Framework" (GBF), including four goals and 23 targets for achievement by 2030.

Its main outcome consists of an agreement, signed by 196 governmental parties, to the effective conservation and management of at least 30% of the world's land, coastal areas and oceans (currently, 17% of land and 8% of marine areas are under protection). It also foresees the restoration of 30% of terrestrial and marine ecosystems, the reduction to near zero of the loss of areas of high biodiversity importance and high ecological integrity. Furthermore it aims at halving global food waste, phasing out or reforming subsidies that harm biodiversity by at least US$ 500 billion per year, while scaling up positive incentives for biodiversity conservation and sustainable use. It should mobilise at least US$ 200 billion per year from public and private sources for biodiversity-related funding and raise international financial flows from developed to developing countries to at least US$ 30 billion per year.

13.8 Conclusion

UNESCO, with its network of around 2000 sites covering almost 6% of the Earth's surface, can play an important role in addressing this challenge by encouraging all of its 195 member states to join the Kunming-Montreal Global Biodiversity Framework. At the same time, more sites should be nominated for UNESCO to achieve 50% of the challenge.

Biosphere Reserves are the ideal places for this challenge with core areas that are legally strictly protected zones and designed to ensure the conservation of landscapes, ecosystems, species, and genetic variation. Buffer zones that surround all core areas to buffer and reduce impacts, are generally used for activities compatible with sound ecological practices that mutually reinforce with scientific research, monitoring, training, and education. Finally, the transition areas, should not be governed by additional regulation but through incentives and innovation, so that communities are empowered to conduct socio-culturally and ecologically sustainable activities, in particular truly Green Economies. Biosphere Reserves include coasts and islands with surrounding marine areas.

References

Clüsener-Godt M, Köck G, Möller L (2022) It's about life: 50 years of UNESCO's man and the biosphere Programme (MAB). Int J Environ Sustain Dev 21(4):377–387

UNEP-WCMC, IUCN, and NGS (2018) Protected Planet Report 2018

UNESCO (2024a) The Zonation of a Biosphere Reserve. https://www.unesco.org/en/mab/wnbr/about

UNESCO (2024b) Infographic, The World Network of Biosphere Reserves 2023–2024. https://en.unesco.org/mab/resources#Infographics

Visual Capitalist (2022) Visualizing carbon storage in Earth's ecosystems, 2022, Dorothy Neufeld, Graphics & Design and Miranda Smith. Data source IPCC, NASA

Zimmer K, (2019) Why the Amazon doesn't really produce 20% of the world's oxygen. National Geographic, Environmental News, August 28, 2019, National Geographic Partners, LLC

Chapter 14
Les Arbres des Océans

Benno Böer, Andrew Dansie, Véronique Helfer, Shreya Kiran, Oleg Shipin, and Martin Zimmer

Abstract This article highlights the growing interest of the international Science-Technology-Innovation community. It points out the potential economic and climatological benefits of a socio-ecological innovation called 'Floating Mangroves', in support of the SDGs 7 (Affordable and Clean Energy), SDG 13 (Climate Action), and SDG 14 (Life Below Water). The authors briefly describe their joint and individual achievements in developing methods to produce mangrove-biomass afloat on tropical and sub-tropical oceans, without the need of terrestrial spaces, and without the need of freshwater supply. They address the urgent necessity of the international scientific and environmental communities to develop this innovation further considering its potential benefits that can contribute to redress the ongoing Triple-Planetary-Syndrom, which is increasingly exacerbated by the global energy paradox.

Keywords Ocean forestry · Floating mangroves · Innovative mangrove development · Clean biofuel · Biomass · Seawater · Clean energy · Freshwater conservation

---

B. Böer (✉)
UNESCO Multi-Sectoral Regional Office, New Delhi, India
e-mail: b.boer@unesco.org

A. Dansie · S. Kiran
University of New South Wales, Sydney, Australia

V. Helfer · M. Zimmer
Leibniz Centre for Tropical Marine Research (ZMT), Bremen, Germany

O. Shipin
Asian Institute of Technology, Bangkok, Thailand

M. Clüsener-Godt et al. (eds.), *Blue Carbon Mangrove Ecosystems*, Blue Carbon Ecosystems for Sustainable Development 1,
https://doi.org/10.1007/978-3-031-69553-7_14

14.1 Introduction

Many compartments of the world are facing massive challenges to supply the population and the economy with energy. Energy poverty has become a defining issue in everyday life.

In view of the ongoing global energy crisis, wood, which is considered a renewable form of energy, offers an alternative to coal, oil and gas, because of its essentially climate-neutral production and use. Wood as a renewable raw material promotes independence from fossil fuels and offers opportunities for a decentralized energy supply. Moreover, wood is a safe raw material, all of which can be used and all of which can be recycled. Even its residues from incineration can be used again, for example as biochar or ash for fertilization or for the production of soap. The long-term use for wood as construction material from sustainable forestry and silviculture can be considered a compensation for hard-to-abate greenhouse gas emissions. The use of wood could therefore contribute even more to climate protection than is already the case.

According to the German Federal Ministry of Food and Agriculture, replacing fossil fuels with wood saves around 32 million tons of CO_2 equivalents every year (BMLE 2023). According to the same source, around 1.1 million households in Germany use firewood (wood pellets; wood chips) as the primary energy source for heating. In addition, there are more than 11 million wood-burning stoves. 65% of the heat from renewable energies in Germany has so far been obtained from wood. The share of wood in the total energy supply in Germany is around 5%. Thus, regenerative energy from wood processing makes an important contribution to the success of the energy transition.

According to 'Die Umweltberatung' (2023), 17% of households in Austria, a country with 46% of its total area covered with forests, are currently heated with wood. A further increase in the share of wood biomass as an energy source makes ecological sense.

14.2 Background

On a global scale, wood is one of the most important energy sources. Today, the importance of wood in this respect is increasing again. In recent years, the increased demand for firewood has made a decisive contribution to improving the economic situation of many forest enterprises, including state-owned and municipal ones. An important advantage of wood is its capacity to provide available energy at all times. Untreated wood that is used as a source of energy can be burned directly, or it can be sold to heating or electricity plants that, with the appropriate technical equipment, can fire this wood as a biological energy source. However, there are natural limits to the growth of the fuel-wood sector.

Fig. 14.1 Mangrove nurseries, like this in Ras al Khaimah can provide thousands of indigenous mangrove saplings for coastal plantation projects. (Photo with kind permission of © Abdulrahman Nasser, edited by B. Böer)

Some have suggested planting and harvesting large amounts of mangroves in inland deserts under sea water irrigation. Besides the fact that mangroves do not need seawater but thrive well in freshwater, we hold this approach as controversial, as this could cause irreversible soil salinization and salinization of fresh groundwater resources, and therefore do not recommend testing this idea.

Others have made laudable efforts with (re-) establishing coastal mangroves, for example in various places in the United Arab Emirates (Fig. 14.1).

14.3 Innovation

By contrast, Böer and Sutcliffe (2014) provided evidence that mangroves, including mangrove wood, can be produced in an innovative technology, floating on top of the world's subtropical and tropical oceans. Such floating mangroves do neither need irrigation with freshwater, nor do they require energy for pumping or drainage, and they do not compete with other land-uses, such as silviculture. The authors developed a small- scale model of floating mangroves. This concept could potentially contribute to the reduction of atmospheric and oceanic carbon levels, reduce land-based marine pollution, generate large quantities of wood-biofuel and other cash crops, and thus generate jobs and income. Numerous globally significant co-benefits

Fig. 14.2 Experimental plot of floating mangroves in Lusail City, Qatar. (Photo: © UNESCO)

would, among others, include enhanced biodiversity, abatement of storm surges and coastal erosion through wave attenuation, seascape improvement, boost to coastal and oceanic fishery stocks, and additional coral-reef support. These potential co-benefits need to be scientifically studied in detail and properly documented.

This short communication offers new insights on the Floating Mangrove concept, based on efforts conducted between 2014 and 2023. It is planned to develop further the existing experiments with floating mangroves and upscale them within technically and environmentally feasible limits. The innovative suggestion to connect floating mangroves to clean energy production and carbon sequestration might add to climate resilience and clean energy security.

A small-scale model of floating mangroves was developed in 2012 (Fig. 14.2), in partnership between UNESCO, Lusail Real Estate Company, and Mourjan Marinas. Seeds of the mangrove species *Avicennia marina* (Forssk.) Vierh. were collected from natural mangrove stands in Qatar, and placed in floating boxes with semi-permeable membranes, allowing for keeping the sandy substrate inside the floating boxes, while allowing sea water intrusion into the root-zone of the mangroves to supply sufficient moisture and nutrient fluxes. The boxes were kept afloat by air-filled plastic containers. In this experiment the seeds germinated and developed into small trees, with a high survival rate until today. They are now growing inside the floating containers, without any utilization of freshwater, as water supply is provided directly by the sea water.

Since 2014, UNESCO has continuously advocated for the further development of a larger scale proto-type of floating mangroves, produced a number of specific proposals, presentations, and a feasibility study. However, until now no funding could be secured, probably because this technology is completely new and therefore no experienced experts can comprehensively review these visionary proposals. Furthermore, the long-term trials needed for the concept validation might deter potential sponsors who frequently look for shorter term returns, despite the fact that mangroves are known to be very efficient blue carbon ecosystems.

Nonetheless, in 2019, the University of New South Wales (UNSW) Water Research Laboratory and Global Water Institute in Australia carried out scientific studies on pontoon design and stability for floating mangrove plantations. An 8.33:1 down-scaled physical model of a 7-m mangrove tree on a 10-metre square pontoon was constructed in a 1.2 m wave flume at the UNSW Water Research Laboratories. The pontoon design was tested both as a floating breakwater structure to attenuate wave energy for stability and with four increasing mangrove tree height and weight scenarios. Two-dimensional testing was undertaken with wave conditions that reflect the expected wave characteristics in sheltered bays. Results were measured using a combination of video footage, accelerometer measurements, capacitance wave probes and visual observations. Wave attenuation and interaction was measured using incident, reflected and transmitted wave heights and used to calculate reflection, transmission and wave dissipation coefficients and wave energy dissipation (Ashley 2019).

The physical model testing demonstrated that the floating mangrove pontoon design was a viable solution in potentially providing sustainable water-based forestry and coastal defence options, particularly under conditions of short wave period.

Further to this, Kiran (2022) assessed the potential of implementing floating mangrove pontoons (FMP) in the Maldives to increase the nation's adaptive capacity to climate change, as a nature-based solution. A total of 14 participants, comprising of experts and professionals involved in mangrove practices at the local, government and international level, were interviewed to assess the potential of establishing FMP in the Maldives. The responses were synthesised using a thematic analysis to identify key recurring themes.

Their results indicate interest among local, government and international stakeholders in establishing a project testing FMP in the Maldives. The findings further identified a range of socioeconomic aspects to be considered, should an FMP project be implemented. Further research is required to understand potential large-scale benefits of FMP to local communities (Kiran 2022).

The Asian Institute of Technology (Thailand), having found the concept interesting and inspirational, also carried out a number of activities related to developing floating mangroves further. One of the key elements of the concept, a floating platform was taken to a promising engineering level. Proposed is an appropriate seaworthy platform of fiber-reinforced foamed concrete. Weighing 3 tons, 1 m high, 3 m in diameter, this platform is very resilient to wave surges and strong winds. It is a walkable platform expected to safely bear mangrove trees for up to 50 years. Numerous single platforms can be flexibly interconnected to form extensive floating tree groves. The most promising platform tree species are thought to be *Rhizophora* and *Avicennia spp.,* because of their wide geographical range and scientifically documented high salinity tolerance.

14.4 Outlook

Based on these new developments, UNESCO managed to identify and motivate reliable and competent partner institutes to try and develop proto-types of floating mangroves, with a view to enhance the availability of wood as clean bio-fuel which, in turn, will assist reducing the energy crisis and contribute to SDG 7 (Clean and Affordable Energy), and to provide additional co-benefits, such as coastal protection and biodiversity-support.

The remaining challenge is to find funding resources, even though the costs to develop a proto-type and required technical, ecological and economic data are estimated at only 2.5 million US$, a very low amount of money in comparison to climate-related catastrophic events, which are in the range of billions of US$.

Considering the extreme freshwater limitations in arid regions, and the need to import and transport fossil fuel over long distances to Small Island Development States (SIDS), where decision-makers and politicians are increasingly concerned with water-, and energy-supply, the large-scale application of floating mangroves to produce wooden biofuel could significantly contribute to solving these issues. UNESCO and UNSW are in discussion of developing the concept further in one of the Pacific SIDS countries.

Inspired by the floating mangrove concept developed by UNESCO, the Leibniz Centre for Tropical Marine Research (ZMT, Bremen, Germany) recently expanded the concept to floating vegetated coastal ecosystems (mangrove stands, saltmarshes and seagrass beds) and included it in a proposal aiming at evaluating the potential of natural and newly designed blue carbon ecosystems for carbon sequestration and long-term storage, including their co-benefits. If successful, the project will test and evaluate the potential of this novel and innovative technology, looking at lateral transfer of organic matter and carbon across ecosystems, and at GHG exchange with the atmosphere and marine waters.

UNESCO is deeply engaged in the monitoring, scientific research and sustainable utilization of mangrove forests, for example, via the inclusion of mangroves in UNESCO-designated sites, such as Biosphere Reserves, World Heritage sites and Global Geoparks, contributing to improving the knowledge, management and conservation of mangrove ecosystems throughout the world. The International Day for the Conservation of the Mangrove Ecosystems, celebrated each year on 26 July, aims to raise awareness of the importance of mangrove forests for humanity. It is very clear that mangrove forests need to be conserved and restored, and more

Fig. 14.3 *Sonneratia* sp. and *Excoecharia* sp. mangroves in the upper intertidal zone in the largest stand-alone closed canopy mangrove forests in the world, in the Bangladesh Sundarbans, illustrating the enormous amount of biomass that mangrove forests produce. (Photo: ©UNESCO/B.Böer 2023)

scientific and development work is required to foster and improve mangrove (re-) establishment (Zimmer et al. 2022).

Mangrove forests are restricted to grow in the upper inter-tidal zone. They do not extend further landward due to competition with other plants, and they do not occur in permanently inundated areas (Fig. 14.3). These restrictions leave just a narrow fringe of suitable geomorphic and sedimentary settings where mangroves can naturally thrive. However, if humankind can develop the floating mangrove concept and make mangroves float on top of the oceans, the narrow range of space that restricts mangrove forests could be significantly increased into the vastness of the world's oceans, with ample supply of space, moisture and nutrients. It goes without saying, though, that in-depth ecological studies are needed to accompany such efforts and make sure that mangrove afforestation or expansion through floats will not negatively affect other valuable coastal or marine ecosystems.

The system is simple, replicable, inexpensive in terms of initial investment and long-term maintenance, and it does not require any energy and equipment for pumping.

New discussions building on the ecological and engineering experiments to date has been initiated between UNESCO, the Asian Institute of Technology (AIT), in Thailand, UNSW, in Australia, and the ZMT in Germany.

In an Op-Ed in 2022, that was jointly released by UNESCO, AIT and UNSW, the three partners agreed '*that further tests are necessary, together with some basic scientific research, to develop this promising new technology in order to obtain science-based data and knowledge for wider and possibly profitable applicability, that can be useful for the production of biofuel (wood-chips, wood-pellets; charcoal)'*. The authors suggested that, '*To be sure, a whole new system of forestry on the ocean's surface could be developed, for the production of fuel... But it is not only for producing useful commodities (...) The carbon-sequestration capacity of these systems needs to be demonstrated, as well as the capacity to take up land-based marine pollutants, in particular nitrogen, phosphates and potassium...UNESCO, AIT and UNSW are in discussion on how best to advance this innovative technology and to demonstrate to the general public and investors this new system for the production of seawater-based trees, that can function as a new and clean source of energy, improve ecosystems services and livelihoods for coastal communities, as well as for the sequestration of atmospheric carbon.*'

The concept was once again presented during the January 2023 Blue Carbon Forum in Yokohama, highlighting the fact that the global Wood Chips market in 2021 stood at 8.8 billion $ and is expected to grow to 13.1 billion $ by 2027, whereas the global Charcoal Market stood at ca. 5 billion $ in 2021 and is expected to climb to ca. 7 billion $ by 2027.

14.5 Conclusions

The most important and pressing questions regarding the implementation of the floating mangrove concept are: '*How much biomass can be produced and used in the form of wood-energy?*'; '*How much carbon can be sequestered?*'; '*What areal extent could be dedicated to those novel designed ecosystems while ensuring no harm to natural coastal ecosystems?*' Other important questions embrace investment costs, design, materials, protection against high wave and wind energy, as well as profitability.

It has been proven that the system works at small scales. We should therefore most certainly develop it further and obtain reliable scientific data to find robust answers to the above questions. Therefore, the development of a proto-type and the collection of scientific data over at least a 5-year period is necessary.

References

Ashley JM (2019) Floating mangrove pontoons, Bachelor of engineering (Environmental engineering). Honours, University of New South Wales

BMLE (2023). https://www.bmel.de/DE/themen/wald/holz/waerme-aus-holz.html (German Federal Ministry of Food and Agriculture)

Böer BCH, Sutcliffe M (2014) Floating mangroves: the solution to reduce atmospheric carbon levels and land-based marine pollution? In: Khan MA, Böer B, Öztürk M, Al Abdessalaam TZ, Clüsener-Godt M, Gul B (eds) 2014: Sabkha ecosystems Vol IV: cash crop halophyte and

biodiversity conservation. Tasks for vegetation science 47. Springer, Dordrecht, pp 327–333. 339p
Die Umweltberatung (2023). https://www.umweltberatung.at/holz-als-energietraeger
Kiran S (2022) Floating mangrove plantations in the Maldives. Bachelor of engineering (Civil and environmental engineering). Honours, University of New South Wales
Zimmer M, Ajonina GN, Amir AA, Cragg S, Crooks S, Dahdouh-Guebas F, Duke N, Fratini S, Friess DA, Helfer V, Huxham M, Kathiresan K, Kodikara KAS, Koedam N, Lee SY, Mangora MM, Primavera J, Satyanarayana B, Yong JWH, Wodehouse D (2022) When nature needs a helping hand: different levels of human intervention for mangrove (re-)establishment. Frontiers in Forests and Global Change 5:784322. https://doi.org/10.3389/ffgc.2022.784322

Chapter 15
ITTO's Cooperation on Sustaining Mangrove Ecosystems

Tetra Yanuariadi

Abstract Mangrove ecosystem is a key contributor to the net-zero carbon emission target the international community is set to achieve by 2050. Mangrove ecosystems and their valuable contributions to sustainable forest management in the tropics, climate change adaptation and mitigation, biodiversity conservation, and enhancement of livelihoods will be a very important area of work for ITTO in the coming years. To develop enabling conditions for sustainable mangrove management, coherent actions at the local, national levels, including research into the causes of depletion and possible solutions, and resource mobilization for large-scale ameliorative action through partnerships and international cooperation need to be in place. ITTO has been working in the area of enhancing mangrove ecosystem development through project implementation since the early years of the organization in the 90 s in three regions: Latin America and the Caribbean, Africa, and Asia-Pacific, as well as at the global level.

Keywords Aquaculture · ITTO · Local community · Sustaining mangrove ecosystem · World mangrove center

15.1 The Need to Sustain Tropical Forests

Globally, forests cover around 31% of the planet's land surface and are essential for life on Earth. Forests are among the world's most productive land-based ecosystems in which an estimated 1.6 billion people depend on forests for subsistence such as food, fuel, fodder and shelter, as well as livelihoods, employment and income.

About 45% of all forests are tropical forests which cover 1.84 billion hectares. They provide a wide range of goods and ecosystem services. Sustainable Forest

T. Yanuariadi (✉)
International Tropical Timber Organization (ITTO), International Organizations Center, Yokohama, Japan
e-mail: tetra@itto.int; https://www.itto.int

M. Clüsener-Godt et al. (eds.), *Blue Carbon Mangrove Ecosystems*, Blue Carbon Ecosystems for Sustainable Development 1,
https://doi.org/10.1007/978-3-031-69553-7_15

Management is a key factor in sustaining tropical forests for the necessities of livelihoods and the environment. When sustainably managed, tropical forests provide vital ecosystem services. The sustainably managed forests can optimally perform their functions to conserve soil and water, regulate climate, keep biodiversity richness, provide fresh air, avoid land degradation and desertification, reduce the risk of floods, landslides, droughts, prevent other disasters, and produce timber and timber products and other non-timber forest products.

Applying SFM's principles and good practices in harvesting and processing tropical timber and other forest products is an important foundation of the tropical forest sector. Developing domestic and international trade for timber and timber will contribute to local and national economies, and gives value to tropical forests, which is essential to reduce forest conversion to agriculture and other land uses.

Deforestation and forest degradation continue at alarming rates and the world lost an estimated ten million ha of forest (an area the size of the Republic of Korea) per year between 2015 and 2020. It is already well known the drivers of tropical deforestation and land degradation include competing land uses with other productive sectors such as agriculture, energy, mining and transportation infrastructure. Land degradation, illegal logging, woodfuel collection, fragmentation, disturbances from insects, disease, severe weather events and wildfires, etc. are major drivers.

ITTO is a treaty-based international organization to promote the sustainable management and conservation of tropical forests and to promote the expansion and diversification of trade in tropical timber from sustainably managed and legally harvested forests. ITTO is the only intergovernmental organization focused exclusively on the sustainable management of tropical forests and the sustainable and legal trade of tropical timber and timber products. ITTO pioneered the concept of sustainable forest management (SFM) in the early 1990s and has been a leader in the development and application of criteria and indicators for SFM. The Organization has also formulated internationally agreed guidelines to support the sustainable management of natural tropical forests, from forest landscape restoration to the conservation of biodiversity in production forests.

The objectives of SFM from ITTO's point of view include:

- To satisfy the needs for goods and services from forests continuously;
- To Ensure efforts to conserve biodiversity;
- To Ensure efforts to conserve forest soils, water and carbon stock;
- To increase resilience and capacity of forests, and carbon storage;
- To support forest-dependent communities for their food security, cultural and livelihood needs; and
- To promote equitable sharing responsibilities in forest management and the benefits arising from forest use.

Since it became operational in 1987, ITTO has funded more than 1200 projects, pre-projects and activities valued at more than USD 430 million, including mangrove projects in several tropical countries of ITTO members. The majority of these have been field projects that have helped operationalize ITTO's policies and guidelines and build capacity in tropical member countries. Many projects have provided multiple benefits, such as in terms of climate-change mitigation, biodiversity conservation, the provision of ecosystem services, disaster risk reduction and community development.

Mangrove ecosystems exist along tropical and subtropical coastlines, distributed worldwide and offer a range of ecosystem benefits on several scales from local to global, including its ability to store 3–5 times more carbon per unit area than any terrestrial ecosystem. The mangrove ecosystem is a key contributor to the net-zero carbon emission target the international community is set to achieve by 2050.

The disappearance trend of the mangrove ecosystems is still at an alarming rate over the past 50 years. Drivers of mangrove deforestation and loss have been well known, namely: i) agriculture expansion (plantation and farmland); ii) aquacultural development (unsustainable fish and shrimp ponds); iii) illegal cutting and land clearing; iv) infrastructural activities (coastal development, urbanization, pollutions etc.); and v) poor management (transparency, law enforcement, and governance) and poor practices of Environmental, Social and Governance (ESG) actions.

15.2 Continuing Mangrove Ecosystem's Functions

Well-managed mangrove ecosystems can support the livelihoods of millions of coastal people while storing globally significant quantities of carbon and reducing the vulnerability of coastal regions to storm surges and other threats. Unfortunately, the areas of mangrove ecosystems are decreasing rapidly in many tropical countries caused of several interlinked factors, such as poor management and increasing demand for coastal lands for development. Therefore, it is a priority for the global community to halt the loss of mangroves and restore degraded mangrove ecosystems through several collaborative efforts:

- Improve policies and regulatory frameworks related to mangrove ecosystem management in accordance with each country's condition, taking into account the existing local wisdom;
- Increase legal access to sustainably utilize non-timber products and environmental services of mangrove resources to improve the livelihoods of local communities;
- Campaign to increase awareness and community engagement in mangrove management and use to maximize their economic, social and environmental benefits;
- Conduct research to provide silvicultural prescriptions and environmentally sound techniques for harvesting mangrove resources; and
- Strengthen collaboration in rehabilitating the mangrove ecosystem —both nationally and internationally.

Mangrove ecosystems are among the most carbon-rich biomes forming highly productive ecotones between terrestrial and marine ecosystems, with carbon production rates equal to those of tropical humid forests. Mangrove forests allocate more carbon below ground than above ground, creating large carbon pools in soils and dead roots that ultimately reach ocean bottoms, away from the atmosphere. Mangrove deposits also enhance the nutritional value of the coastal waters, which become breeding grounds for marine life.

Mangrove conservation fits well into the global effort to pursue the SDGs targets and fulfill the Paris Agreement on climate change. Mangroves could play a crucial role in implementing the Paris Agreement on climate change by reducing the vulnerability of coastal regions to the changing climate and because of their high potential to remove carbon from the atmosphere and thereby contribute to mitigation. Achieving mangrove conservation requires placing it at the forefront of governmental conservation priorities and carrying out sustained awareness-raising campaigns to convince people of the seriousness of the situation and the need for urgent action. Non-consumptive economic uses of mangroves on a significant scale would help reduce pressure. Good mangrove governance is underpinned by responsive policies, strategic planning, the active restoration of degraded sites, the zoning of production and protection areas, curtailing the conversion of mangrove areas for other uses, and active local participation and capacity building.

15.2.1 Facing Continued Threats

Drivers of mangrove ecosystem losses and degradation have changed in recent years, with rapidly developing economies requiring large investments in coastal infrastructure. This has reduced the effectiveness of some important societal responses towards conservation and sustainable management. Mangroves are under serious threat worldwide due to over-exploitation and poor resource management, conversion for infrastructural uses such as ports, resorts and roads as well as for aquaculture and rice cultivation, and marine pollution. A lack of clarity on tenure rights has contributed to the weakening of community's role in mangrove protection.

The management of mangrove forests requires integrated multi-sectoral and multi-disciplinary approaches, as well as better coordination and networking. Societal and governmental responses to the fast-deteriorating situation of mangroves have been inadequate. Community-based mangrove management, which has traditionally been confined to providing organizational inputs and labor, is unable to cope with the requirements of capital investments. Planned land-use change for other economic activities is another major cause of mangrove loss, and a lack of awareness of the true value of mangroves and a lack of coordination among various authorities also places serious stress on this precious resource.

15.2.2 Effective Local Community Participation

The starting point for restoring degraded mangrove areas should be the preparation of integrated management plans based on an area's bio-geophysical background and environmental history. These plans should be prepared by qualified mangrove professionals with the full collaboration of local communities and local governments, and they should incorporate the development of viable alternative livelihoods for mangrove-dependent families. Involving local communities and adopting local agendas would be beneficial not only for ecological sustainability but also for the well-being of communities that depend on coastal resources.

Community-based forest management approach in a mangrove ecosystem can succeed when there is clarity on tenure; attractive immediate and long-term incentives exist; capacity is developed; and there is good local awareness of the role of mangroves and sound knowledge of appropriate restoration techniques. To ensure that communities extend their consent for restoration overall restoration sites, they should be provided with appropriately thought-out exit strategies that limit damage to the restored ecosystems. In places where local people profit commercially from mangroves, user fees could be incorporated for financing the management, rehabilitation and conservation of mangroves.

Non-consumptive uses of mangroves for ecotourism and nature education can help reduce resource dependency and should be promoted. Even with good investment in capacity building, however, it takes time to develop areas for ecotourism because it is not easy to change the attitudes of local communities. Being an entrepreneurial activity, ecotourism involves risks that local communities need to understand.

15.3 Making the Difference on the Ground

Mangrove ecosystems and their valuable contributions to sustainable forest management in the tropics, climate change adaptation and mitigation, biodiversity conservation, and enhancement of livelihoods will be a very important area of work for ITTO in the coming years. ITTO has been working in the area of enhancing mangrove ecosystem development through project implementation since the early years of the organization in the 90 s in three regions: Latin America and the Caribbean, Africa, and Asia-Pacific, as well as at the global level. Those projects were presented in the boxes below.

Box 15.1 ITTO's Mangrove Project Implemented in the Latin America and Caribbean region

1. Management and Conservation of Mangroves in the Gulf of Fonseca, Honduras
2. Evaluation of Mangrove Forests in the North-East of the Orinoco Delta Region in Venezuela
3. Sustainable Management and Rehabilitation of Mangrove Forests by Local Communities on the Caribbean Coast of Colombia

(continued)

Box 15.1 (continued)

4. Management, Conservation, and Development of Mangrove Forests in Panama
5. Criteria for the Management of Mangrove and Flood Forests in the Central Coastal Plains of Veracruz, Mexico
6. Environmental Assessment and Economic Valuation of Ecosystem Services Provided by Coastal Forests in the Central Coastal Plains of Veracruz, Mexico
7. Strengthening Mangrove Ecosystem Conservation in the Biosphere Reserve of Northwestern, Peru
8. Strengthening of Governance and Sustainable Management of Mangrove Ecosystem in Guatemala as a Climate Change Adaptation Measure

Box 15.2 ITTO's Mangrove Project Implemented in the Africa Region

1. Assessment and Management of Mangrove Forests in Egypt for Sustainable Utilization and Development
2. Study for the Conservation, Rehabilitation and Sustainable Management of Mangroves in Togo
3. Sustainable Community Management, utilization and Conservation of Mangrove Ecosystems in Ghana
4. Regeneration and Management of Mangrove Forests Surrounding the Douala/Edea Reserve, Cameroon

Box 15.3 ITTO's Mangrove Project Implemented in the Asia-Pacific Region

1. Establishment of an International Network for the Conservation and Sustainable Utilization of Mangrove Forest Genetic Resources, in India
2. Community-Based Restoration and Sustainable Management of Vulnerable Forests of the Rewa Delta, Viti Levu, Fiji
3. Achieving Sustainable Management of Mangrove Forests in China through Local Capacity Building and Community Development
4. Demonstration of Integrated Models for Sustainable Management of Mangrove in China
5. Assessment of Mangrove Forest Affected by Cyclone Nargis to Facilitate the Development of an Integrated Mangrove Ecosystem Management in Ayeyarwady Delta, Myanmar
6. Model of Rehabilitation and Sustainable Utilization of Mangrove Forest at Langkat, North Sumatra, Indonesia
7. Promoting Local Community Initiatives on the Rehabilitation of the Mangrove Ecosystem in Bintan Island, Indonesia

15.4 Strengthening Cooperation to Sustain Mangrove Ecosystems

To develop enabling conditions for sustainable mangrove management, coherent actions at the local, national levels, including research into the causes of depletion and possible solutions, and resource mobilization for large-scale ameliorative action through partnerships and international cooperation need to be in place.

In 2017, ITTO in collaboration with the Indonesian Ministry of Forestry and the International Society for Mangrove Ecosystems (ISME) convened an International Conference on Sustainable Mangrove Ecosystems: Managing a Vital Resource for Achieving the Sustainable Development Goals and the Paris Agreement, in Bali Indonesia (ITTO 2017). The conference sent messages that are still applicable in recent situations to mangrove development stakeholders at local and global levels:

- Prioritizing the conservation and sustainable management and use of mangrove ecosystems in national policies, laws and regulations at all levels of government, and strengthening law enforcement to reduce the continuing loss and degradation of mangrove ecosystems;
- Promoting sound land-use planning, clarifying land tenure and user rights for communities with consideration for cultural sensitivity and ensuring the effective involvement and empowerment of local communities, especially women;
- Promoting effective mangrove restoration through the improved assessment of degraded and deforested mangrove ecosystems, science-based decision-making to identify the best restoration strategies and practices, and long-term monitoring and management to determine and improve the overall success of restoration programs and increase productivity;
- Facilitating access to international financing mechanisms as well as national public and private sector to scale up mangrove-based climate-change mitigation and adaptation and enhance the resilience of coastal regions and communities; and
- Generating and disseminating knowledge on the cause, patterns and consequences of change in mangrove ecosystems at the local to global scales, reporting on the status of mangrove management worldwide, and raising awareness of the vital importance of mangrove ecosystems to planetary well-being.

The most recent activity on ITTO cooperation with other international organizations to sustain the mangrove ecosystem was to support the development of the World Mangrove Center initiated by the Government of Indonesia and the Government of Germany. Led by CIFOR (Center for International Forestry Research), ITTO and the Leibniz Centre for Tropical Marine Research (ZMT) acted as the Advisory Team and produced a feasibility study for the development of the center (CIFOR, ITTO, ZMT (2022).

Being home to the largest area of mangroves worldwide, Indonesia is well suited to having a prominent research, learning, training and communication center, from which new knowledge may be generated. The center will, in turn, share knowledge and experiences with other like-minded countries for the betterment of unique mangrove ecosystems currently facing tremendous pressures. The goal of the center is to deliver transformative and action-oriented evidence and solutions to addressing the key drivers of mangrove deforestation and loss. As an international hub for knowledge and cooperation that promotes sustainable management and stops the conversion of mangroves, the center will bring together international mangrove researchers and stakeholders to advance our understanding, develop solutions and share knowledge to bring it into application to support the healthy mangrove ecosystems and mangrove dependent communities in light of the climate crises.

The World Mangrove Center was launched by the President of Indonesia, Joko Widodo, on 10 June 2022 (ITTO 2022).

References

CIFOR, ITTO, ZMT (2022) Establishment of the world mangrove center (WMC). An advisory Team's report was submitted to the Centre for Sustainable Forestry Management Instrument Standardization, Ministry of Environment and Forestry of the Republic of Indonesia

ITTO (2017) Report of international conference on sustainable mangrove ecosystems: managing a vital resource for achieving the sustainable development goals and the Paris agreement. Prepared by the Natural Resources Development Centre (NRDC). The conference was held on 18–21 April 2017 in Bali, Indonesia

ITTO (2022) ITTO Strategic Action Plan 2022–2026. Policy Development Series 25

ITTO Website (www.itto.int): provides varieties of information related to ITTO's mission, activities and projects

Chapter 16
Blue Carbon Activities in the Pangatalan Island Within Palawan Biosphere Reserve, The Philippines

Frédéric Tardieu and Laure Thierry de Ville d'Avray

Abstract The Sulubaaï Foundation, established on the island of Pangatalan within the Palawan Biosphere Reserve, in the Philippines, is actively involved in the conservation and restoration of Blue Carbon Ecosystems, focussing on protecting and rehabilitating mangrove forests, seagrass beds, and coral reefs, which are vital habitats for marine life and play a crucial role in carbon sequestration. This article provides an outline of the activities of the foundation related to blue carbon in the island of Pangatalan (The Philippines).

Keywords The Sulubaaï foundation · Pangatalan · Palawan Biosphere reserve · The Philippines · Blue carbon ecosystems · Mangrove forests · Seagrass beds · Coral reefs · Marine Life

16.1 Introduction

Since its creation in 2011, the Sulubaaï Foundation has gradually developed increasingly extensive and community-based actions for the restoration of ecosystems. Pangatalan Island has been and remains the pilot site for the actions of the Foundation which are intended to extend to the whole bay and to be replicated in other sites in collaboration with the communities (Fig. 16.1).

The Foundation conducts various activities to achieve their goals. These include mangrove reforestation efforts, by planting new mangrove trees to restore degraded areas. They also engage in seagrass bed restoration projects, aiming to enhance the health and resilience of these important ecosystems (Fig. 16.2).

F. Tardieu (✉)
President Sulubaaï Foundation, Pangatalan, Philippines

L. T. de Ville d'Avray
Researcher at Sulubaaï Foundation, Pangatalan, Philippines

M. Clüsener-Godt et al. (eds.), *Blue Carbon Mangrove Ecosystems*, Blue Carbon Ecosystems for Sustainable Development 1,
https://doi.org/10.1007/978-3-031-69553-7_16

Fig. 16.1 The Island of Pangatalan in the Palawan Biosphere Reserve. (Copyright Sulubaaï 2023)

Fig. 16.2 Mangroves on Pangatalan Island. (Copyright Sulubaaï 2023)

Additionally, the Sulubaaï Foundation conducts research and monitoring programs to better understand the ecological dynamics of Blue Carbon Ecosystems. The Foundation collaborates with local communities, scientists, and other stakeholders to raise awareness about the importance of these ecosystems and promote sustainable practices.

16.2 Key Activities/Outcomes

It's worth noting that the Foundation's efforts contribute not only to the conservation of marine biodiversity but also to climate change mitigation. Blue Carbon Ecosystems have the ability to capture and store significant amounts of carbon dioxide, thus help in combating the global warming.

Before the creation of the Foundation, Pangatalan was a devastated island, the vegetation had been cut down for wood and the mangroves for making coal (management plan 2016/2021). The main goal was the transformation of this deserted island into a model of sustainable development where eco-tourism could contribute to environmental restoration. The Foundation team was entirely composed of workers from the surrounding villages. Its goal was to develop a sustainable model that would be integrated into local communities.

Since 2015, the Sulu-Reef-Prosthesis has been created for coral reef restoration, an artificial reef design, and many Marine Protected Areas were created. The activities of the Foundation include a marine ecosystems diagnosis with the help of marine biologists. The findings, which were relevant to the entire bay, confirmed that the coral reefs were heavily impacted by destructive activities such as dynamite fishing. In that sense, the Foundation developed a restoration technique adapted to this environmental context, which respects biodiversity, is easy to set up, very inexpensive and can be replicated. Regular monitoring is carried out in order to assess the efficiency of the coral restoration method, and results are meant to be published in peer-reviewed scientific journals.

In March 2016, a 46-hectare marine protected area (MPA) surrounding the island has been created by the Sulubaaï Foundation, in agreement with the barangays (villages) of Depla & Silanga, as well as the municipality of Taytay, see Fig. 16.3. The MPA has a highly diverse marine environment where at least 13 ecosystems have been identified.

In 2021, the Sea Academy was created for sustainable management of marine resources with local communities in order to restore ecosystems. As Pangatalan has become a model of sustainable development and ecosystem restoration, the Foundation wanted to extend its measures to the entire bay. The local communities that had been interacting with the Foundation for years wanted to collaborate, convinced by the successful results. The Sea Academy programme is a scientific platform and educational programme that was created with local communities (see Fig. 16.4) and our international partners. It consists in creating a network of 4 marine protected areas managed in collaboration with local communities and is based on four principles: Protection, Restoration, Education and Replication. The results will be published in scientific publications, network communications, and international news reports.

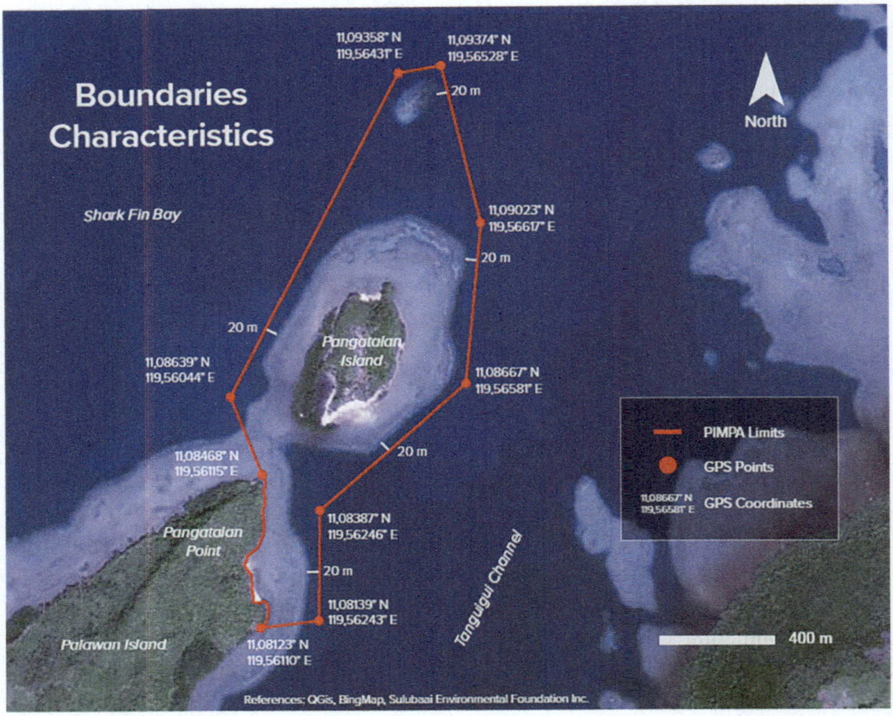

Fig. 16.3 Pangatalan marine protected area

16.3 Conclusion

Overall, the Sulubaaï Foundation of Pangatalan is dedicated to the protection and restoration of Blue Carbon Ecosystems in the Philippines, and their activities are aimed at preserving these valuable habitats for future generations.

Fig. 16.4 Local communities interacting with the Sulubaaï Foundation. (Copyright Sulubaaï 2023)

Reference

For further information, see also the website of the Sulubaaï Foundation: https://sulubaai-foundation.com/our-pangatalan-island/

Chapter 17
Mangrove Blue Carbon Ecosystems in Saudi Arabia

Ronald A. Loughland

Abstract Saudi Arabia has extensive coastlines, occurring both along the Saudi Gulf within the temperate zone, and along the Red Sea in the tropics and temperate zone. The prevailing conditions within these climatic zones greatly influences mangrove species distribution, growth form and density, with the tropical zone being optimal for mangrove growth.

In the southern Red Sea within the tropical zone, two species of mangrove (*Avicennia marina* and *Rhizophora mucronata*) occur, and their well-developed growth forms and higher density reflect the optimal growing conditions at that latitude. In the northern temperate areas of the Red Sea and along the entire Saudi Gulf coastline, only one species of mangrove persists which is *A. marina*. *A. marina* growing at these locations exhibits growth forms and density typical of mangrove habitats occurring at latitudinal extremes, such as low heights and sparse distribution.

The remaining natural mangrove habitats along the Saudi Gulf coastline are now confined to isolated remnant patches within embayment's, or on nearshore islands, with over 90% of Gulf mangrove habitats having been lost due to coastal development. Along the Red Sea coastline, mangrove habitats remain mostly unaffected by development, with camel grazing being the main threat. However as Red Sea coastal development increases, it's imperative that the mangrove habitats are protected, and the country via the National Center for Vegetation Cover and Combating Desertification (NCVC) embarked on a major initiative to increase vegetation cover, including the protection of remnant mangroves within new national parks and the plantation of millions of new mangroves.

The mangroves of Saudi Arabia are significant blue carbon ecosystems, sequestering and storing large volumes of carbon annually. These intertidal habitats provide climate change adaptation for the country, along with ecological, community and food security benefits. Carbon quantification and blue carbon project registration has commenced in Saudi Arabia, and with the country's ambitious

R. A. Loughland (✉)
UNDP, Riyadh, Saudi Arabia

M. Clüsener-Godt et al. (eds.), *Blue Carbon Mangrove Ecosystems*, Blue Carbon Ecosystems for Sustainable Development 1,
https://doi.org/10.1007/978-3-031-69553-7_17

re-vegetation plans, mangroves and other blue carbon habitats are set to play an important role in assisting Saudi Arabia meet its carbon neutrality targets by 2060.

Keywords Shallow water marine habitats · Blue carbon ecosystems in Arabia · Saudi Arabian mangrove forests · Mangrove plantation initiatives in the Arabian Gulf and Red Sea · Blue carbon accreditation process in the Arabian region · Historic decline of mangrove forest in the Arabian Gulf and their restoration · Blue carbon value for the conservation and restoration of coastal vegetation communities in Arabia

17.1 Mangrove Distribution, Growth Characteristics and Species Dispersal

Mangroves occur mostly within the tropical zone (Fig. 17.1). Mangrove ecosystems are intertidal communities occurring mostly in sheltered locations along warm coastal shorelines. These communities often form dense forests, characterized by various species of mangrove, that can be stratified in species specific bands from the coast—inland. In the tropics optimal mangrove growing conditions occur including high rainfall and fresh water runoff, yearly warm weather and an abundance of nutrients. The tallest and most dense mangrove ecosystems grow in the tropics,

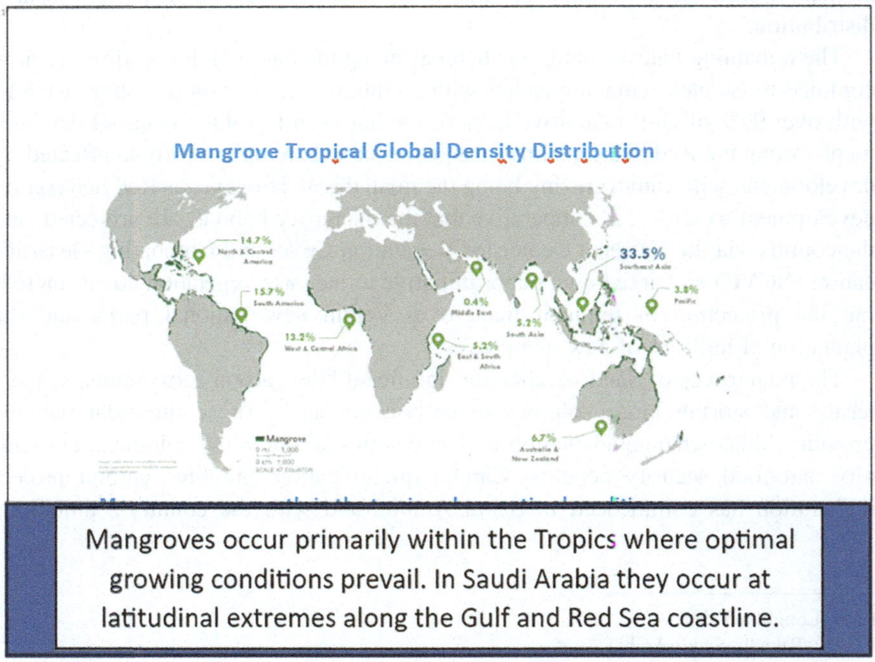

Fig. 17.1 World distribution of Mangroves

where they accumulate more biomass and sediments, meaning that they sequester and store much more Carbon than mangrove communities that occur outside the tropics at latitudinal extremities, outside optimal climatic conditions. At these extreme latitudes mangroves often form monocultures (Saenger 2002) and can become dwarfed and occur as scattered shrubs (Saenger and Loughland 2019).

Mangroves in Saudi Arabia comprise some of the most northern latitudinal limits of the Indo-Pacific-East-African mangroves (Al-Mahasheer 2018), with only one species of mangrove *Avicennia marina* (Forsk.) Vierh. surviving within the Gulf and two species (*A. marina* and *Ryzophora mucronata*) occurring in the Red Sea, with *A. marina* occurring along the entire Red Sea coastline and R. mucronata only in the warmer southern part of the Red Sea (Saenger and Loughland 2019). *A. marina* is tolerant of relatively high salinity (> 40 ppt) together with low rainfall and extreme temperature fluctuations. This species characterizes the high latitude limits of mangroves in the Arabian Peninsula. Mangrove communities at their latitudinal limits are generally characterized by decreasing plant height and by their patchy distribution (Dodd et al. 1999; Saenger and Loughland 2019). Both characteristics are apparent in the mangrove populations along the northern Saudi Arabian coastline of the Gulf and Red Sea. For example, mangrove growth heights on the shores of the Saudi Gulf rarely exceed 5 m, with most trees occurring around 1.5 m in height (Saenger and Loughland 2019) reflecting the harsh environment (high salinity, low rainfall, extreme temperature fluctuations) in which they grow.

On the Saudi Arabian shores of the Gulf, only three widely separated natural mangrove communities remain, all confined to highly sheltered embayment areas (Saenger and Loughland 2019). The largest natural mangrove forests remain within Tarut Bay and are predominately on the northern shore of Tarut Island and within the coastal area between Rahima and Safwa (Rahima Embayment). These remaining mangrove forest fragments constitute less than 10% of the original mangrove cover in the Bay.

Figure 17.2 illustrates the influence climate has on the growth of mangrove. The same species *Avicennia marina* grows throughout the global mangrove range and is the only species occurring along the Gulf coast of Saudi Arabia. At different latitudes from the equator the growth form of this species varies, and ranges from tall robust trees in Asia on the equator, to medium sized trees in the southern tropical areas of the Red Sea, and stunted trees in the harsher temperate climate of the Gulf. The height within this species ranges widely from 30 m in Asia to 1.5 m in the Gulf, and this is influenced primarily by climatic conditions.

Along the northern Red Sea coast of Saudi Arabia occurs an isolated natural population of *A. marina* that is growing at latitudinal extremes. The population survives in a geomorphological refugia and is in stark structural contrast to *A. marina* populations growing in the southern (tropical) Red Sea coast of Saudi Arabia.

In the southern tropical zone of the Saudi Red Sea, south of Yanbu, both *A. marina* and *R. mucronate* form dense mangrove forests, where species zonation is evident, particularly in the very south near Jizan (Fig. 17.3). The narrow coastline and adjacent high elevations of the escarpment in this region ensures that adequate

Fig. 17.2 *A. marina* growth form at different locations

Fig. 17.3 Zonation of Mangroves

freshwater runoff occurs within coastal wadi systems, primarily in the summer months, with freshwater discharge along the Red Sea coastline into mangrove habitats. In this warm moist environment mangroves thrive, protected from wave energy by the fringing coral reefs that occur along the Red Sea coastline. The conditions in the southern Red Sea are therefore optimal for mangrove development, with dense and tall mangroves forests occurring within sheltered embayment's.

17.2 Threats to Mangroves in Saudi Arabia

Mangrove communities along the Gulf shoreline of Saudi Arabia and the nearby island of Bahrain have declined significantly over the last 50 years (Zayani and Loughland 2009; Loughland and Al-Abdulkader 2011; Al-Abdulkader and Loughland 2011; Al-Abdulkader et al. 2019). Along the Gulf coast of Saudi Arabia, Loughland et al. (2011) documented that more than 90% of mangroves had been lost as a result of sprawling coastal development, and alarmingly most of the loss had occurred recently within the last decades. The same study identified and prioritized mangrove habitats at imminent risk from development and in need of urgent protection.

Mangrove communities along the vast Red Sea coast of Saudi Arabia were until recently little impacted by coastal development, as coastal communities occurring outside of the larger population centers were sparsely distributed and typically small. The main threat to mangrove ecosystems along the Red Sea is from livestock grazing, especially large herds of camels that graze directly on the vegetation (leaves, flowers and seeds) of mangroves throughout the entire year.

In recent years, large coastal tourism, and real estate developments have commenced along the Red Sea coast, with the pristine nature of the coastline (and its mangroves) being the main attraction for the developments. With the correct environmental guidelines in place and their strict enforcement, the mangrove communities of the Red Sea should continue to persist and flourish.

17.3 Ecological and Economic Value of Mangroves in Saudi Arabia

Mangroves provide a resilient dynamic vegetative buffer between terrestrial and marine ecosystems and are therefore important for impeding coastal erosion and buffering storm and tidal surges; filtering sediments, nutrients and pollutants from terrestrial runoff; and providing a complex structured habitat for species both above, between and below water (e.g., migratory and resident birds and bats, coastal wading birds, fish, shrimp, crabs and invertebrates). Mangrove ecosystems are also very efficient at removing and storing atmospheric CO_2, hence providing a valuable nature-based solution for climate change. Mangrove ecosystems are very productive supporting entire marine food chains through the export of nutrients into adjacent marine ecosystems such as seagrass and coral reefs.

With the significant loss of mangrove habitat along the Saudi Gulf coastline, the last few remaining native mangrove communities have immense ecological and biodiversity value. These remnant forests play a critical role as nurseries for fisheries, that many local communities depend on to support their livelihoods. The continued existence, and expansion of these forests supports fundamental food security for the Kingdom.

17.4 Mangrove Plantations, Past, Present and Future

Provided with optimal germination and growing conditions, mangrove will colonize intertidal shorelines and become established. These optimal conditions include (1) warm air temperature and high humidity within the mangrove tolerance range, (2) muddy or sandy substrate, (3) protection from wave energy, (4) suitable tidal range, and (5) shallow gently sloping shorelines (Saenger and Loughland 2019). In order to expand mangroves beyond areas where these optimal conditions prevail, careful site selection and engineering efforts are required that create suitable growing conditions where mangrove seedlings can survive, mature and reproduce. This has been successfully undertaken within the extreme latitudes of the Gulf (Loughland et al. 2020) and has proven to be a feasible solution to restore mangrove habitats in areas that have had the original conditions altered from development activities; or in natural coastal areas that have low ecological and economic value such as coastal sabkha (coastal saline deserts) that can be converted into productive mangrove ecosystems (Loughland and Saenger 2001; Loughland et al. 2013; Loughland et al. 2014).

There are mangrove plantation programs being undertaken in Saudi Arabia, and these commenced initially along the Gulf shoreline, and were latter expanded also along the Saudi Red Sea coastline. The first plantations occurred in the early 1990s within Khore Khafji in the northern Gulf, where exotic species of mangrove were established (*A. germinans, R. mucronata*), with *A. germinans* still surviving. Following the ecological impacts of the Gulf war's massive oil spill, efforts to plant native *A. marina* at other locations along the Gulf commenced. The then Saudi Wildlife Authority, and the King Fahad University for Petroleum and Minerals (KFUPM) undertook a number of trial plantations along the Gulf coastline commencing in 1993. The most successful of these was on the island of Abu Ali near Jubail where less than 100 planted seedlings developed over three decades into 14 hectares of dense mangrove forest. There were additional mangrove plantation programs conducted throughout the 1990s and into the twenty-first century by the Saudi Arabian Oil Company (Saudi Aramco) in cooperation with the then Ministry of Fisheries and Agriculture and KFUPM, and these were small scale plantations with the main focus being on raising awareness amongst local school and community groups. The Royal Commission for Yanbu and Jubail also commenced similar small scale mangrove plantations at their respective coastal locations during this same period.

Following the publication of Saudi Arabia's first Marine Atlas (Loughland and Al-Abdulkader 2011) that documented the sharp decline of over 90% of mangroves along the Gulf coastline, the author initiated with the support of Saudi Aramco a systematic mangrove restoration program at a much larger scale along the shoreline of the Saudi Arabian Gulf. This resulted in the establishment of strategic nurseries along the Gulf coast, where seed collection from remnant forests, and a seedling hardening (acclimatization) process was implemented by gradually increasing sea-water salinity to reduce stress during periods of low temperatures. This resulted in significant increased seedling survival (98%) and the successful planting out of millions of acclimatized mangroves at suitable locations along the Gulf coastline from 2011–2020. The program incorporated for the first-time hydrological surveys of potential plantation sites to identify and mark-out the correct tidal zonation for planting, ensuring that the seedlings would flourish within the optimal tidal cycle range (Loughland and Saenger 2001).

In 2020, the author technically developed and led Saudi Aramco's Green Energy Program, that supported a much larger scale mangrove plantation program across the entire Kingdom (Gulf and Red Sea) and at tropical locations globally. The objective of this program was to develop blue carbon nature-based solutions for atmospheric carbon sequestration and storage. This program was fundamental to assist Saudi Aramco meet its Net Zero Target by 2050, with the Green Energy Program aiming to remove up to 30% of total company unavoidable emissions.

The Kingdom of Saudi Arabia has also recently focused on mangroves as a blue carbon nature-based solution for meeting its Net Zero Target of 2060. In association with the Kingdom's Vision 2030, and the Saudi Green initiative, the Kingdom aims to plant over 10 billion trees, with many of these being mangroves, because water scarcity, hyper-arid conditions and high costs of maintaining terrestrial trees would be unsustainable. In support of the National Vision, many private Saudi corporations are now engaged in planting trees and mangroves in their operating areas under corporate stewardship programs coordinated through the National Center for Vegetation Cover and Combating Desertification (NCVC).

The NCVC was established in Riyadh under the Ministry of Environment, Water and Agriculture. The Center is focusing on protecting mangrove remnant habitats through a series of national parks and planting new mangroves at suitable coastal sites along both the Gulf and Red Sea coastlines. The NCVC will coordinate all future mangrove plantations across the Kingdom, because with the recent interest of various sectors to plant millions of mangroves, a national strategy was developed by NCVC. The strategy identifies the correct locations, and ensures suitable seed is sourced from genetically correct populations. The strategy also ensures that planting programs are implemented using the appropriate methodologies to maximum survival of planted seedlings.

17.5 Mangrove Conservation, Increasing Awareness and Changing Opinions

Following the publication of the Marine Atlas outlining the rapid decline in mangrove habitats along the Gulf coastline of Saudi Arabia (Loughland and Al-Abdulkader 2011), the conservation of the remaining remnant mangrove habitats became a priority. Along the Gulf coastline a 64 km^2 blue carbon ecosystem centered on Rahima Bay, part of the larger Tarut Embayment was incorporated as a corporate biodiversity protection area by Saudi Aramco in 2012 including remnant mangrove forests, saltmarsh, seagrass and sabkha habitats (Fig. 17.4). The Marine Atlas authors worked on a design concept for the Kingdom's first Mangrove Eco Park facility within the reserve inclusive of a mangrove boardwalk. Saudi Aramco implemented the design and developed the Rahima Mangrove Eco Park, which was officially opened in January 2021. The Eco Park is open to the public and receives hundreds of visitors each year. The facility provides awareness information on mangrove and blue carbon ecosystems and their important ecological and climate change role. For the first time in the Kingdom, the public can easily enter the intertidal mangrove world via a 500 m boardwalk. At the facility carbon sequestration is monitored using Eddy Flux Covariance towers, and a dedicated laboratory analysis carbon storage within intertidal sediments. This information is then displayed to visitors in the Center.

In 2021, UNESCO registered the Kingdom's first Man and the Biosphere Reserve in the Red Sea on the Farasan Islands primarily for the conservation of *R. mucronata*

Fig. 17.4 The Tarut Bay Mangrove Eco Park

Fig. 17.5 *R. mucronata* growing on limestone outcrops on Farasan Islands, Southern Red Sea

populations that grow on limestone structures in the Red Sea (Fig. 17.5). Following this the NCVC embarked on planning for the establishment of mangrove national parks around the Kingdom.

There has been a fundamental shift in public opinion regarding mangroves in the Kingdom. Within the last decade the importance of mangrove habitats has been recognized by the Government, Corporations and especially the younger generations, with social media assisting in the dissemination of information. The previous decline of mangroves has halted, and through restoration programs mangrove habitats are now increasing.

17.6 Mangrove Blue Carbon Research and Development

In a study by the author (Cusack et al. 2018) carbon stocks in surface sediments were measured across three blue carbon habitats along the Gulf coast of Saudi Arabia, with comparable stocks for saltmarsh (81 ± 22 Mg Corg ha^{-1}), seagrass (76 ± 20 Mg Corg ha^{-1}) and mangroves (76 ± 23 Mg Corg ha^{-1}). For mangroves, there was a 38% decrease in carbon stocks between mature established mangrove stands (91 Mg Corg ha^{-1}) and planted mangroves (56 Mg Corg ha^{-1}). Mangrove habitats were the most efficient carbon sequesters, with the highest carbon burial rate of the three ecosystems (19 g Corg m^{-2} $year^{-1}$), compared to seagrass (9 g Corg m^{-2} $year^{-1}$) and saltmarshes (8 g Corg m^{-2} $year^{-1}$).

In 2022 Saudi Aramco's mangrove plantations on Abu Ali Island within the Gulf were surveyed for below and above ground carbon storage as part of a carbon credit verification program conducted with an international verifier.

This was the very first blue carbon credit project undertaken in the Kingdom and was coordinated by the author as part of Saudi Aramco's Green Energy Program to secure carbon credits for offsetting emissions. The carbon credit project was based on the Kyoto Protocol, where carbon credits could be used by organizations to meet their emission reduction targets.

In order to meet all requirements, the plantation's carbon credits needed to be validated and certified by a carbon credit verifier registry. Registries have specific methodologies and protocols, particularly related to plantations. The fundamental principles included land ownership, documented objectives of utilizing the project for obtaining carbon certification, confirmation of commitment for long-term land/plantation protection and monitoring, engagement with local communities, and plans for involvement throughout the entire crediting period (carbon credit monitoring normally occurs each 2–3 years and can continue up to 60 years). It was also a requirement to document the land conditions and any previous management (baseline) prior to the commencement of the project, along with the project planting program, calculations of potential GHG emissions whilst developing the program and the project safeguards that ensure long-term protection of the plants.

Once all the project requirements were met, a comprehensive project description document was developed. The project description underwent public review for 30 days before being validated and verified by the Validation and Verification Body (VVB). The VVB conducted a site visit to verify the data and information documented in the project description and submitted a final validation report to the registry for registration and carbon credit issuance.

17.7 Conclusion

Saudi Arabia has two coast lines, the Gulf and the Red Sea. The Red Sea coastline is extensive and ranges from the tropics in the south to a temperate cool climate in the north. The Gulf is situated out of the tropics within the temperate zone, and the climate occurring along both coastlines influences mangrove species distribution, growth form and density.

A.marina is the only species of mangrove growing along the coastline of the Saudi Gulf coast, and constitutes some of the most northern mangrove habitats in the world. *A. marina* growth form and density is greatly influenced by the climate prevailing at these extreme mangrove latitudes. Similarly, along the Red Sea coastline, *A. marina* is the only mangrove occulting along the northern temperate coastline, and shares the coast with *R. mucronate*a in the southern tropical coastal areas of the Red Sea, where distinct species zonation is apparent.

There has been a significant reduction in mangrove habitats along the Saudi Gulf Coast due to coastal developments. Less than 10% of the original mangrove habitat still remains, and these remnants are now protected and are being expanded via plantation programs. Along the Red Sea coast, mangrove habitats are in good condition, and have not been greatly impacted in the past from coastal development, with the main threat being camel grazing. The future protection of these mangrove habitats is important as coastal development is now increasing rapidly along the Red Sea coastline.

Mangrove habitats in Saudi Arabia are significant Blue Carbon sinks, sequestering and storing large volumes of carbon annually, particularly in the southern tropical zone of the Red Sea. These same blue carbon intertidal habitats also ensure long-term adaptation to climate change whilst providing valuable ecological and community benefits. The quantification and registering of mangrove carbon credits has commenced in Saudi Arabia, and with ambitious plans to restore vegetation cover, particularly mangroves, there is huge potential for Blue Carbon Ecosystem expansion and monitoring in Saudi Arabia.

References

Al-Abdulkader K, Loughland RA (2011) Conservation of marine and coastal habitats. In: Loughland RA, Al-Abdulkader K (eds) . Natural Resource Atlas of the Western Arabian Gulf, Saudi Aramco, Kingdom of Saudi Arabia, p 455

Al-Abdulkader K, Loughland RA, Qurban MA (2019) Ecosystems and biodiversity of the Arabian gulf, Saudi Arabian waters—fifty years of scientific research. Saudi Aramco, Verona, p 354

Al-Mahasheer H (2018) Spatial coverage of mangrove communities in the Arabian Gulf. Environ Monit Assess 190:85

Cusack M, Saderne AV, Masqué DP, Krishnakumar K, Rabaoui L, Qurban M, Qasem AM, Prihartato PK, Loughland RA, Duarte CM (2018) Carbon sequestration and storage of blue carbon within ecosystems in Saudi Aramco protected wetlands. Environ Res Lett

Dodd RS, Blasco F, Rafii ZA, Torquebiau E (1999) Mangroves of The United Arab Emirates: ecotypic diversity in cuticular waxes at the bioclimatic extreme. Aquat Bot 63:291–304

Loughland RA, Saenger P (2001) Report to the Public Works Department of Abu Dhabi on the methods for the development of mangrove plantations in Abu Dhabi Emirate. Department of Environmental Research, Emirates Heritage Authority, Abu Dhabi

Loughland RA, Al-Abdulkader K (eds) (2011) Natural resource Atlas of the Western Arabian Gulf. Saudi Aramco, Kingdom of Saudi Arabia, p 455

Loughland RA, Wyllie A, Al-Abdulkader K (2011) Anthropogenic induced changes along the Gulf coast of KSA from 1967–2010. In: Piacentinj T (ed) Geomorphology. Intech, Rijeka, p 333

Loughland RA Qasam, A, Burwell B (2013) Coastal Sabkha (salt flats) of the southern and Western Arabian Gulf. In Encyclopedia of wetlands: wetlands of the world (vol. 4) coastal Sabkha. Ecosystems book-springer, p177

Loughland RA, Qasam A, Burwell B (2014) Research and development with seawater and halophytic plants for sustainable saline agro systems in the Arabian Gulf. In: Khan MA et al (eds) Sabkha ecosystems: volume IV: cash crop halophyte and biodiversity conservation, tasks for vegetation science 47. Springer Science Media, Dordrecht

Loughland RA, Butt SJ, Nithyanandan M (2020) Establishment of mangrove ecosystems in Kuwait: sustainable outcomes in a challenging and changing environment. J Aquat Bot 167: 103273

Saenger P (2002) Mangrove ecology, silviculture and conservation. Kluwer Academic, Dordrecht

Saenger P, Loughland RA (2019) Mangroves and saltmarsh. In: Al-Abdulkader K, Loughland RA, Qurban MA (eds) Ecosystems and biodiversity of the Arabian gulf, Saudi Arabian waters—fifty years of scientific research. Saudi Aramco, Verona, p 354

Zayani AK, Loughland RA (2009) Intertidal habitats. In: Loughland RA, Zainal AJ (eds) Natural resource atlas of Bahrain. GEOMATEC, Manama, pp 85–113

Chapter 18
Blue Carbon and the Concept of "Coastal Wetlands" in the IPCC 2013 Wetlands Supplement

Hans Joosten

Abstract Various criteria to define blue carbon ecosystems were discussed with the definition of "coastal wetlands" in the IPCC 2013 Wetlands Supplement. This paper gives an overview of these considerations.

Keywords IPCC guidelines · UNFCCC reporting · Peatland · Brackish · Astronomical tides · Restoration

18.1 Introduction

According to the US National Oceanic and Atmospheric Administration (NOAA), *blue carbon* is "carbon captured by the world's ocean and coastal ecosystems" (https://oceanservice.noaa.gov/facts/bluecarbon.html). The IPCC sixth Assessment Report defines *blue carbon* as "biologically driven carbon fluxes and storage in marine systems that are amenable to management" (IPCC 2021). The report notices, that "coastal blue carbon focuses on rooted vegetation in the coastal zone, such as tidal marshes, mangroves and seagrasses." The report furthermore observes that "there is current debate regarding the application of the blue carbon concept to other coastal and non-coastal processes and ecosystems, including the open ocean" (IPCC 2021).

These descriptions raise questions about the concrete features, either individually or in combination, that characterize and define blue carbon ecosystems:

- Must blue carbon ecosystems be wetlands?
- Must they be saline/brackish?
- Must they be close to the sea and how close?

H. Joosten (✉)
Institute of Botany and Landscape Ecology, Greifswald University, partner in the Greifswald Mire Centre, Greifswald, Germany
e-mail: joosten@uni-greifswald.de

M. Clüsener-Godt et al. (eds.), *Blue Carbon Mangrove Ecosystems*, Blue Carbon Ecosystems for Sustainable Development 1,
https://doi.org/10.1007/978-3-031-69553-7_18

- Must they be influenced by (lunar) tides?
- Must they be rich in carbon? And how rich?

Similar questions were addressed in the definition of *coastal wetlands* in the 2013 Wetlands Supplement (IPCC 2014). This paper gives an overview of that Supplement and these considerations.

18.2 The 2013 IPCC Wetlands Supplement

Recognizing the problem of global climate change, the World Meteorological Organization (WMO) and the United Nations Environment Programme (UNEP) established in 1988 the Intergovernmental Panel on Climate Change (IPCC). One of IPCC's activities is to develop methodologies ('Guidelines') to assist countries in compiling national inventories of greenhouse gases (GHGs) for reporting to the UN Framework Convention on Climate Change (UNFCCC).

The "2013 Supplement to the 2006 IPCC Guidelines for National Greenhouse Gas Inventories: Wetlands" (in short "Wetlands Supplement") has been produced in response to an invitation from the Subsidiary Body for Scientific and Technological Advice (SBSTA) of the UNFCCC (December 2010). SBSTA at its 33rd Session in Cancun (Document FCCC/SBSTA/2010/L.18, paragraph 4) invited IPCC "to undertake further methodological work on wetlands, focusing on the rewetting and restoration of peatland, with a view to filling in the gaps in the 2006 IPCC Guidelines for National Greenhouse Gas Inventories". This guidance had become necessary because agreement had been reached among UNFCCC negotiators on including peatland rewetting as a new accountable activity in the Kyoto Protocol (Joosten 2010). For reporting on that activity, insufficient guidance was available in the existing 2006 IPCC Guidelines (IPCC 2006).

The Wetlands Supplement has taken its mandate broadly and provides guidance for reporting on inland drained and rewetted organic soils, mineral soil wetlands, coastal wetlands including mangrove forests, tidal marshes and seagrass meadows, and constructed wetlands for wastewater treatment (IPCC 2014). The Supplement starts with a decision tree guiding the user to the respective chapters or back to the 2006 Guidelines in case guidance is not provided by the Supplement (Table 18.1).

In spite of its confusing name, the Wetlands Supplement is not restricted to the IPCC/UNFCCC land use category "Wetlands" but addresses emissions and removals from wetlands and drained soils in any land use category, i.e. in Forest Land, Cropland, Grassland, Wetlands, Settlements and Other Land. The land-use category under which land is concretely reported depends on national definitions and approaches. For example, plantations on peatland may be classified as Forest Land or as Cropland depending on the national forest definition, and mangroves may be classified as Forest Land or as Wetlands.

The guidance should allow any country, regardless of experience or resources, to produce reliable estimates of its greenhouse gas emissions and removals (IPCC

Table 18.1 Decision aide to determine where GHG reporting guidance on organic soils and various wetland use types is to be found. (Modified after IPCC 2014)

1	Is the soil used for rice cultivation?	Yes	2006 IPCC Guidelines and Chaps. 2 and 5 of the 2013 Wetlands Supplement
		No	→ 2
2	Is the soil organic and/or wet?	No	2006 Guidelines
		Yes	→ 3
3	Is this a "flooded land"?	Yes	2006 Guidelines and the 2019 Refinement (IPCC 2019)
		No	→ 4
4	Is this a wetland for waste-water treatment	Yes	Chapter 6 of the 2013 Wetlands Supplement
		No	→ 5
5	Is this coastal land?	Yes	Chapter 4 of the 2013 Wetlands Supplement
		No	→ 6
6	Is this inland mineral soil?	Yes	Chapter 6 of the 2013 Wetlands Supplement
		No	→ 7
7	Is this organic soil wet?	Yes	Chapter 3 of the 2013 Wetlands Supplement
		No	Chapter 2 of the 2013 Wetlands Supplement

2014). No guidance is, however, provided for saline inland wetlands (playas, pans, salt lakes, brackish wetlands, salinas, sabkhas), as too little information was available in the literature to assess potential greenhouse gas emissions or removals from these lands.

The Wetlands Supplement covers land with organic soils, and wet and drained mineral soils across all land-use categories. Within these land-use categories, the Supplement differentiates between coastal land and inland land, because water salinity and dynamics (e.g. tides) may modify emission factors compared to inland land. The Supplement proposes a division into four soil subcategories (drained mineral soil, wet mineral soil, wet organic soil, and drained organic soil), each with a coastal and inland subdivision where appropriate (see Fig. 18.1).

The Wetlands Supplement classification system is strictly dichotomous: Every soil that is not an organic soil is classified as a mineral soil, every soil that is not a wet soil is a dry soil, all land that is not coastal is inland.

18.3 Wet, Drained and Rewetted

The Supplement defines a *wet* soil as a soil that is inundated or saturated by water for all or part of the year to the extent that biota, adapted to anaerobic conditions, particularly soil microbes and rooted plants, control the quality and quantity of the net annual greenhouse gas emissions and removals.

Drainage is defined as the process of artificial lowering of the soil water table. In the Supplement, the term is used to describe the act of changing a soil from wet into dry. A *drained* soil is a soil that has formerly been a wet soil, but as a result of human

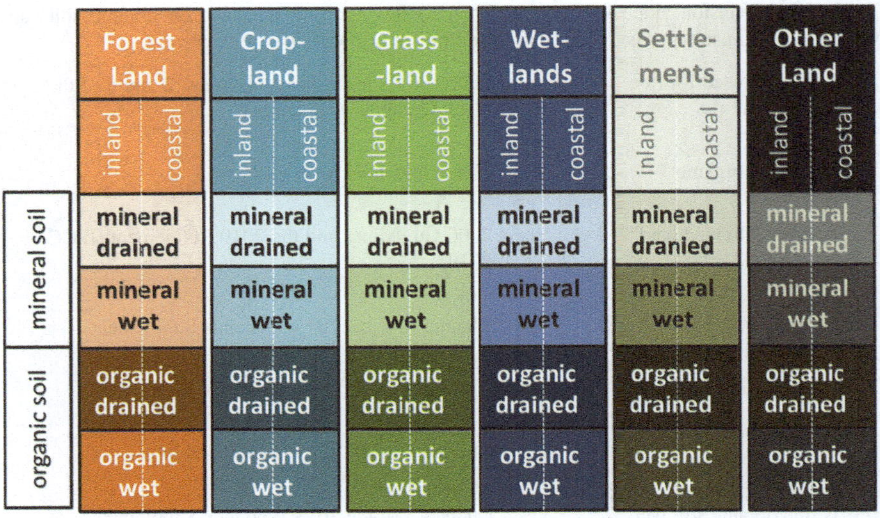

Fig. 18.1 Soil based land-use subcategories addressed in the 2013 Wetlands Supplement. (IPCC 2014)

intervention has become a dry soil. All organic soils are assumed to have originally been wet, therefore a dry organic soil is always also a drained organic soil. Rewetting is the process of changing a drained soil into a wet soil. A *rewetted* soil is a soil that has formerly been a drained soil but as a result of human intervention has again become a wet soil. Restoration (adjective restored) is the process of assisting the recovery of an ecosystem that has been degraded, damaged, or destroyed. In the case of drained former wetlands, restoration always has to include rewetting.

18.4 Coastal Wetlands

The Supplement defines *coastal land* as land at or near the coast. All land that is not coastal is inland. Countries are expected to clearly define the concept of "coastal land" and its sea- and landward limits in accordance with their national circumstances, and to apply that definition consistently across the entire national land area and over time.

A *coastal wetland* is defined as a wetland at or near the coast that is influenced by brackish/saline water and/or astronomical tides. Brackish/saline water is water that normally contains more than 0.5 or more parts per thousand (ppt) of dissolved salts. The boundary of coastal wetlands may extend to the landward extent of tidal inundation and may extend seaward to the maximum depth of vascular plant vegetation. Coastal wetlands may occur on both mineral and organic soils, meaning that they—provided that they are influenced by brackish/saline water or astronomical tides—may also be peatlands.

Countries need to develop a nationally appropriate definition of coastal wetland taking into account national circumstances and capabilities.

Coastal wetlands can potentially occur in any land-use category. They may by management either remain in the same category or shift to another land use category. Seagrass meadows classified as Wetlands, for example, may remain reported as Wetlands following introduction of aquaculture or may be reclassified as Settlements. Tidal marshes are classified as Wetlands but after drainage for agriculture they are classified as Cropland or Grassland. Mangroves classified as Forest Land that undergo harvesting or biomass clearing may remain reported as Forest Land if their land use remains the same or be reclassified to another land-use category in case they undergo clearing and drainage.

Three properties thus define a coastal wetland according to the Wetlands Supplement:

1. Being at or near the coast
2. Being influenced by brackish or saline water
3. Being influenced by astronomical tides.

The first property is a *conditio sine qua non*: even if a wetland is saline or influenced by astronomical tides, it is no coastal wetland if it is located too far from the coast. Being influenced by brackish/saline water or by astronomical tides, on the other hand, are necessary but not sufficient requirements. The *sine qua non* means that not all areas that are influenced by astronomical tides have to be included in the concept of coastal wetlands: the tidal influence of the Amazon, for example, may range up to 1000 km from the ocean (Kosuth et al. 2009), i.e. a bit far to still be called "coastal". The same accounts for brackish/saline waters: the *sine qua non* criterion will certainly exclude the brackish Lake Issyk-Kul in the midst of Central Asia's Tien Shan mountains in Kyrgyzstan on an elevation of 1600 m (Vollmer et al. 2002). But as said before: the "at or near the coast" criterion is not strictly fixed: Countries may define the concept themselves, provided that they apply the definition consistently across the entire national land area and over time.

On the other hand, peatlands that do not comply with the provision of being influenced by brackish water or astronomical tides are not included in the concept of coastal wetlands, even not when they are directly bordering the sea (cf. the raised bog near Archangelsk, Russia, and the *Cladium* mire near Lagoa St. André, Portugal, directly bordering the Baltic Sea and the Atlantic Ocean, respectively, and depicted on pictures 29 and 51 in Succow and Joosten 2001).

Let's illustrate the concept with an example from my backyard, the Karrendorfer Wiesen near Greifswald, Germany (Fig. 18.2):

- The Karrendorfer Wiesen are coastal: the open Baltic Sea is visible in the background.
- They are regularly flooded and therefore wetlands.
- They are flooded with brackish water from the Baltic Sea.
- They are flooded by wind tides, especially in autumn and winter, but not by astronomical (lunar) tides, because the Baltic Sea is too small to have own

Fig. 18.2 The Karrendorfer Wiesen near Greifswald, Germany. Courtesy: Tobias Dahms

significant tides, and it has a too narrow opening to the North Sea to be influenced by the North Atlantic tides.

- They are subject to intensive carbon sequestration research (Jurasinski et al. 2020).
- They are coastal, brackish and regularly flooded and they sequester and store peat. In Germany we call them "Küstenüberflutungsmoore" (coastal floodwater mires, our equivalent to peatforming mangroves; Joosten et al. 2017).

References

IPCC (2006) 2006 IPCC guidelines for National Greenhouse gas Inventories. In: Eggelston S, Buendia L, Miwa K, Ngara T, Tanabe K (eds) Institute for Global Environmental Strategies (IGES) for the IPCC, Hayama

IPCC (2014) In: Hiraishi T, Krug T, Tanabe K, Srivastava N, Baasansuren J, Fukuda M, Troxler TG (eds) 2013 Supplement to the 2006 IPCC guidelines for National Greenhouse Gas Inventories: wetlands. IPCC, Geneva

IPCC (2019) 2019 refinement to the 2006 IPCC guidelines for National Greenhouse Gas Inventories. In: Calvo Buendia E, Tanabe K, Kranjc A, Baasansuren J, Fukuda M, Ngarize S, Osako A, Pyrozhenko Y, Shermanau P, Federici S (eds) . IPCC, Geneva

IPCC (2021) Annex VII: Glossary. Matthews JBR, Möller V, van Diemen R, Fuglestvedt JS, Masson-Delmotte V, Méndez C, Semenov S, Reisinger A (eds). In: Masson-Delmotte V, Zhai P, Pirani A, Connors SL, Péan C, Berger S, Caud N, Chen Y, Goldfarb L, Gomis MI, Huang M, Leitzell K, Lonnoy E, Matthews JBR, Maycock TK, Waterfield T, Yelekçi O, Yu R, Zhou B (eds) Climate change 2021: The physical science basis. Contribution of working group I to the sixth assessment report of the intergovernmental panel on climate change. Cambridge University Press, Cambridge, UK/New York, NY, pp. 2215–2256

Joosten H (2010) Getting peatlands under Kyoto: arriving at a moving target in Cancun. IMCG Newsletter 2010(3–4):13–19

Joosten H, Moen A, Couwenberg J, Tanneberger F (2017) Mire diversity in Europe: mire and peatland types. In: Joosten H, Tanneberger F, Moen A (eds) Mires and peatlands of Europe—status, distribution and conservation. Schweizerbart Science Publishers, Stuttgart, pp 5–64

Jurasinski G, Ahmad S, Anadon-Rosell A, Berendt J, Beyer F, Bill R, Blume-Werry G, Couwenberg J, Günther A, Joosten H, Köbsch F, Köhn D, Koldrack N, Kreyling J, Leinweber P, Lennartz B, Liu H, Michaelis D, Mrotzek A, Negassa W, Schenk S, Schmacka F, Schwieger S, Smiljanic M, Tanneberger F, Urich T, Wang H, Weil M, Wilmking M, Wrage-Mönnig N (2020) From understanding to sustainable use of peatlands: the WETSCAPES approach. Soil Syst 4(14):1–27

Kosuth P, Callède J, Laraque A, Filizola N, Guyot JL, Seyler P, Fritsch JM, Guimarães V (2009) Sea-tide effects on flows in the lower reaches of the Amazon River. Hydrol Process 22:3141–3150

Succow M, Joosten H (eds) (2001) Landschaftsökologische Moorkunde, 2nd edn. Schweizerbart, Stuttgart

Vollmer MK, Weiss RF, Williams RT, Falkner KK, Qiu X, Ralph EA, Romanovsky VV (2002) Physical and chemical properties of the waters of saline lakes and their importance for deep-water renewal: Lake Issyk-Kul, Kyrgyzstan. Geochim Cosmochim Acta 66:4235–4246

Chapter 19
Blue Carbon Stocks of Mangrove Conservation Areas in Southeast Asia: Opportunities and Challenges in Forest Protection and Restoration

Dixon T. Gevaña, Carlos Mico M. Villanueva, and Josephine E. Garcia

Abstract One of today's most serious global problems is climate change. Deforestation account to as much as 20% of this problem due to carbon emissions from cutting down trees and its corresponding ecosystem degradation. Forest restoration is vital to reverse this trend. Among the tropical forests, mangroves provide a vital role in mitigating climate change. They offer diverse benefits and services that contribute to human well-being. One of its astonishing services is their ability to greatly sequester atmospheric carbon, which is popularly termed as blue carbon. Global mangrove cover continues to decline at an alarming rate. In Southeast Asia, loss in mangrove cover accounted to 58% from the 1990s to 2015, whereas 30% was mainly due to the conversion of mangrove areas to aquaculture. Other drivers of mangrove loss in Southeast Asia include conversion to rice and oil palm plantations, and coastal reclamation. Serious mangrove loss has prompted the need for adopting conservation programs like those under UNESCO's Man and Biosphere, Ramsar Wetlands of International Importance, and blue carbon project initiatives that are in-lined with REDD Plus. These programs generally aims at better protection of biodiversity,

D. T. Gevaña (✉)
Department of Social Forestry and Forest Governance, College of Forestry and Natural Resources, University of the Philippines Los Baños, Laguna, Philippines

Forestry Development Center, College of Forestry and Natural Resources, University of the Philippines Los Baños, Laguna, Philippines
e-mail: dtgevana@up.edu.ph

C. M. M. Villanueva
Department of Social Forestry and Forest Governance, College of Forestry and Natural Resources, University of the Philippines Los Baños, Laguna, Philippines
e-mail: cvillanueva@up.edu.ph

J. E. Garcia
College of Agriculture and Food Science, Institute of Animal Science, University of Philippines Los Baños, Laguna, Philippines
e-mail: jegarcia4@up.edu.ph

M. Clüsener-Godt et al. (eds.), *Blue Carbon Mangrove Ecosystems*, Blue Carbon Ecosystems for Sustainable Development 1,
https://doi.org/10.1007/978-3-031-69553-7_19

promotion of sustainable livelihoods, and conservation of carbon stocks. This chapter showcases the vital role of Southeast Asian mangroves under various UN's conservation programs in blue carbon storage. Estimated carbon stocks from several studies in selected biosphere reserves, Ramsar sites and protected areas were listed and described. Key findings showed that blue carbon stocks range from 77 Mg C ha^{-1} to 1220 Mg C ha^{-1}, of which a larger part was observed in the sediment pool. As blue carbon studies emerged over the last decade, there exists differences and variations in carbon stock estimation due to varying methodologies in estimation. Overall, mangroves under UN's conservation programs have huge potential for developing blue carbon projects. There are however various biophysical and socio-institutional challenges that need to be understood and overcome. It is necessary to craft local integrated mangrove management plans, adopt collaborative mangrove governance approach, and forge regional alliances and mangrove road map for the ASEAN in order to effectively use blue carbon programs.

Keywords Climate change · Coastal development · Collaboration · Integrated plan · Restoration

19.1 Introduction

Climate change is becoming more pronounced as seen from drastic increases in temperature and sea level rise, loss of glaciers, changes in rainfall patterns, and prolonged periods of drought (Adedeji et al. 2014). Globally, the rise in temperature accounted to an average increase of 0.08 °C every 10 years since the 1980s and is expected to increase for up to 5.4 °C by the end of the twenty-first century. (NOAA 2020; Heshmati 2020). Coastal vegetations such as mangroves play a vital role to mitigate climate change as they help sequester large amount of atmospheric carbon in their vegetation and sediment.

Mangroves are coastal ecosystems that thrive highly in brackish waters and extreme environmental conditions (Kathiresan and Bingham 2001; Giri et al. 2011). They provide diverse goods and services to humans and other organisms. For example, they are sources of wood and non-wood materials, provide protection from tsunamis and coastal erosion, and serve as nurseries and habitats for small fish and many animals (FAO 2005; Mcleod and Salm 2006; McKee 1996; Nagelkerken et al. 2008). Among the ecosystem services, carbon sequestration is underscored to be one of the most important role that mangroves perform (Alongi 2012). However, mangrove cover continues to decline globally at an alarming rate of 21,000 ha per year (FAO 2020). This translates to 1.04 million ha loss between 1990 and 2020. Chowdhury et al. (2017), identified three main drivers of mangrove loss common in many studies, namely: coastal development, mangrove conversion, and wood extraction. Of these three drivers, the conversion of mangroves to aquaculture and agriculture was found to be the most common in Southeast Asia (Primavera 2005a; Richards and Friess 2016; Goldberg et al. 2020). Pendleton et al. (2012) asserted that annual carbon emissions brought by conversions in mangroves is huge at

0.09 to 0.45 Pg CO_2, while 0.0.15 to 1.02 Pg CO_2 for tidal marsh and seagrasses. This shares about 20% of yearly carbon loss due to global deforestation.

Mangrove deforestation can also be attributed to governance problems. The limited policies to protect mangroves, tenurial issues, limited funds for mangrove conservation, and unclear institutional jurisdictions particularly in the government which often favors conversion of mangrove to other land uses, are some of the important issues needing attention (Pulhin et al. 2017; Gevaña et al. 2018a; Ruzol et al. 2020). Further, the limited efforts to empower coastal communities and stakeholders towards collaborative mangrove conservation contribute indirectly to the degradation and decline of mangrove cover (Jentoft 2005; Garnett et al. 2007; Fisher 2014; Gevaña et al. 2021). In response to these, there are several global initiatives that are being pursued to reverse this trend. These include the Global Mangrove Alliance, UNESCO's Man and Biosphere Programme, Ramsar Site nominations, and REDD+ scheme under the Paris Agreement, to name a few.

This chapter endeavors to highlight the need for conserving mangroves of Southeast Asia. It looks at their vital climate change mitigation role specifically blue carbon storage and sequestration, which is one of the major reasons for prioritizing them in restoration and protection efforts. This paper synthesizes the blue carbon estimates from selected biosphere reserves, Ramsar sites, and other protected mangroves in the region. Further, it enumerates and describes the factors contributing to the changes in carbon stocks; and identify challenges in blue carbon initiatives and mangrove management.

The carbon values presented in this study are based on secondary research and multiple literatures related to the blue carbon in Southeast Asian mangroves. These featured literatures were found to be content-authentic and carefully studied and considered important in this study. Therefore carbon stock values were interpreted and collated to better understand and compare mangrove sites in Southeast Asia under unique circumstances, challenges, and management approaches. Furthermore, the evidence, analysis and discussions made in this study to assess changes in carbon estimates and carbon management challenges reflect several examples, site histories, and policies highlighted in several related literature.

19.2 Mangroves of Southeast Asia

Southeast Asia (SEA) consists of approximately 35.6% of the world's mangrove cover. They share the greatest plant diversity representing 51 of the world's known 73 species (Spalding et al. 2010; Giri et al. 2011; Thomas et al. 2017) (Fig. 19.1). According to FAO Global Forest Resources Assessment (2015), Indonesia has the largest area of mangrove cover at 2.24 million ha (56% of the total SEA area) (Fig. 19.2). Then, followed by Malaysia (521,000 ha) and the Philippines (356,000 ha). Timor Leste and Singapore cover only about 3000 ha.

Table 19.1 summarizes the comparison data of mangrove area and percentage of annual rate of change from FAO (2015) and Bunting et al. (2018). Recording the

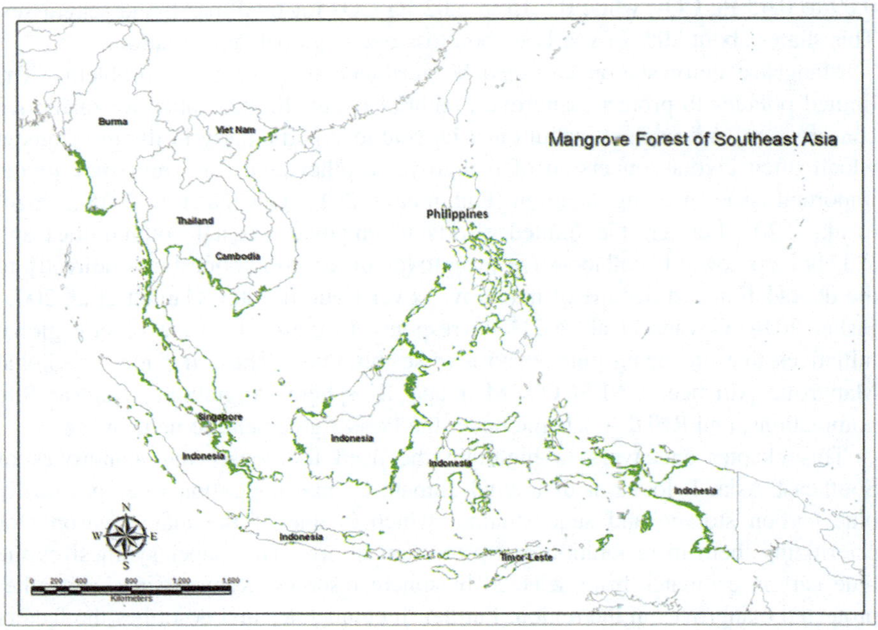

Fig. 19.1 Map of Southeast Asia mangrove forests as shown in dark green color. (Source: Bunting et al. 2018)

Fig. 19.2 Pie chart of mangrove areas in Southeast Asia. (Source: FAO 2015)

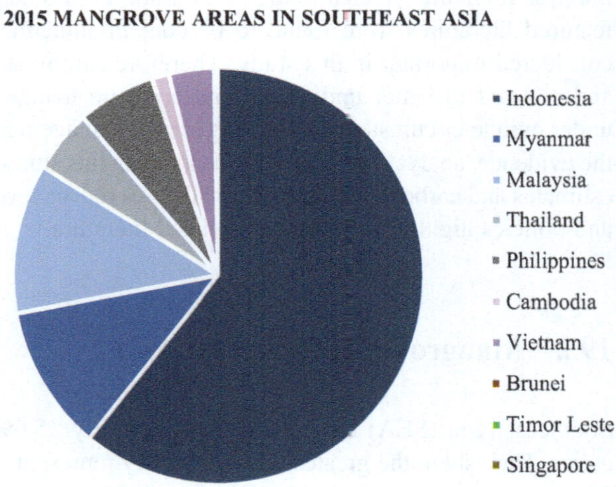

world's total mangrove area as part of FAO's Global Forest Resources Assessment, Southeast Asia sustains around 4.1 million hectares in 2015. The annual rate of change (1990 to 2015) showed a negative trend in mostly SEA countries (e.g., Myanmar, Cambodia, Indonesia), however, Vietnam, Philippines and Thailand presented a significant increases in mangrove area.

Table 19.1 Comparison of mangrove areas, percentage, cover change and annual rate of change in Southeast Asia based on FAO (2015) and Bunting et al. (2018)

Country	Area (FAO 2015)					Annual change of rate (%)	Area (Bunting et al. 2018)				Annual change of rate (%)
	1990	2000	2005	2010	2015		1996	2007	2010	2015	
Indonesia	3058	3062	3444	2844	2244	−1.23	2824	2699	2672	2650	−0.33
Myanmar	517	486	438	438	299	−2.17	538	501	498	492	−0.48
Malaysia	548	529	545	529	521	−0.20	522	511	517	509	−0.13
Thailand	174	245	265	244	240	1.29	236	222	224	213	−0.56
Philippines	154	227	265	311	356	3.41	282	271	264	268	−0.26
Cambodia	80	68	62	56	50	−1.86	63	59	59	58	−0.46
Vietnam	73	71	62	262	270	5.36	171	161	159	157	−0.45
Brunei	18	18	18	18	18	0.00	11	11	11	11	−0.14
Timor Leste	3	2	2	2	2	−1.61	1	1	1	1	−0.21
Singapore	1	1	1	1	1	0.00	1	1	1	1	−0.11
Southeast Asia	4626	4709	5102	4705	4001	−0.58	4649	4435	4405	4539	−0.34

Meanwhile, Bunting et al. (2018) generated a global mangrove baseline map using ALOS, PALSAR and Landsat data, as well as baseline changes from JERS-1, ALOS, and ALOS-2 over various periods from 1996 to 2016. Compared to FAO (2015), the GMW data has a much higher coverage area in 2015 with almost 4.4 million hectares. Conversely, in terms of annual change rate shown in Table 19.1, all SEA countries have negative values (e.g., Thailand (−56%), Myanmar (−48%), and Cambodia (−46%).

Overall, according to FAO and Bunting et al. (2018) the area of mangroves in SEA shows a decreasing trend of 0.58% and 0.34% annual loss, respectively. Southeast Asia has a long history of mangrove exploitation and management, with forest reserves established by colonial administrators in the nineteenth and early twentieth centuries for timber and charcoal production (Friess et al. 2019). In Myanmar, mangroves are used for fuelwood and charcoal production due to the high calorific value of their wood (Oo 2002). Hamilton and Casey (2016) estimated the annual mangrove forest loss averaging 0.26–0.66% globally between 2000 and 2012. The spatial distribution and are of mangrove loss during the twenty-first century is far from even, with rates in Southeast Asia consistently twice as global rate. Myanmar experienced the highest rates of loss at 0.5–0.7% per year from 2000–2012 (Richards and Friess 2016; Hamilton and Casey 2016), more than four times the global mangrove average.

Commodity production is one of the main drivers of mangrove loss in the region. Aquaculture still contributes to 30% of the loss of mangrove forest areas, as well as the conversion to other land uses, such as rice fields in the region (22%, mainly in Myanmar) and oil palm ((Richards and Friess 2016). Even though rates of mangrove loss are currently lower than during the past century, considerable progress is still needed to fully protect and rehabilitate mangrove resources.

19.3 Mangrove Blue Carbon Stock in Selected Ramsar Sites, UNESCO Biosphere Reserves, and Protected Areas in Southeast Asia

Pursuing studies to measure mangrove blue carbon have become vital in understanding the needed efforts towards climate change resiliency. Mangroves play a huge role in climate change mitigation by serving as effective carbon sinks. With this in mind, restoring and protecting mangroves have become a major strategy to respond to adverse climate change. Table 19.2 summarizes the potential carbon stocks of mangroves in selected RAMSAR sites, UNESCO Biosphere Reserves, and other protected areas of Southeast Asia. These sites differ markedly in estimated carbon stocks, as each site follows different mangrove management schemes and several standards set by the Ramsar Convention on Wetlands and UNESCO. Further, the methods used and carbon pools assessed (eg. vegetation; sediment; or both) in the assessments vary among sites which likely contributed to the differences in the

Table 19.2 Blue carbon stock estimates in selected Ramsar sites, UNESCO Biosphere Reserves, and other protected mangrove forests in Southeast Asia

Country	Site	Vegetation type	Area (ha)	Est C –stock (Mg C ha^{-1})	Year of assessment	Evaluated Carbon Pool	Estimation method	References
Vietnam	Mui Ca Mau National Park (RAMSAR) / (UNESCO)	Natural—mixed spp.	13, 400	768.7	2012	Vegetation + Sediment	Non-destructive transect sampling	Tue et al. (2014)
	Can Gio Mangrove Forest Park (UNESCO)	restored and regenerated	44,696 ha	910.7	2016	Vegetation + Sediment	Non-destructive transect sampling	Dung et al. (2016)
	Kien Giang Province (UNESCO)	Young restored	not indicated	76.9	2010	Vegetation + Sediment	plot-based / rapid field assessment	Wilson (2010)
Thailand	Phuket—Phang Nga Bay	Natural—mixed spp.	not indicated	166.3	2019	vegetation	Remote Sensing	Darmawan et al. (2019)
	Krabi Estuary (RAMSAR) (Koh Klang)	Natural—mixed spp.	not indicated	1029.5	2015	Vegetation + Sediment	Non-destructive transect sampling	Elwin et al. (2019)
Malaysia	Pulau Kukup (RAMSAR)	Bruguiera-Rhizophora mangrove forest	647.2	427.9	2016	Vegetation + Sediment	plot sampling	Tan et al. (2012) and Rozainah et al. (2018)
	Tanjung Piai (RAMSAR)	Bruguiera-Rhizophora mangrove forest	526		2016	Vegetation + Sediment	plot sampling	
Indonesia	Sembilang National Park (RAMSAR)	Natural—mixed spp.	87,000	141.0	2011	Vegetation	Non-destructive transect sampling	Manuri et al. (2011)
	Tanjung Puting National Park (RAMSAR) / (UNESCO)	Natural—mixed spp.	11,000	1220.2	2009	Vegetation + Sediment	Non-destructive transect sampling	Murdiyarso et al. (2009)

(continued)

Table 19.2 (continued)

Country	Site	Vegetation type	Area (ha)	Est C –stock (Mg C ha^{-1})	Year of assessment	Evaluated Carbon Pool	Estimation method	References
Philippines	Bahile, Puerto Princesa, Palawan (UNESCO)	Natural—mixed spp.	Not indicated	529.9	2014	Vegetation + Sediment	Non-destructive quadrat sampling	Abino et al. (2014)
	Katunggan It Ibajay (KII) Eco Park	Natural—mixed spp.	44.22	82.1	2017	Vegetation	Line transect method	Barrientos and Apolonio (2018)
Cambodia	Koh Kong / Peam Krasaop WildlifeSanctuary	Natural—mixed spp.	23,750	953.9	2019	Vegetation + Sediment	Plot sampling	Sharma et al. (2019)
Singapore	Pulau Ubin	Natural—mixed spp.	249.3	226.9	2014	Vegetation + Sediment	Remote sensing,	Friess et al. (2015)
	Sungei Buloh Wetland Reserve	Natural—mixed spp.	238.2	163.7	2014	Vegetation + Sediment	Remote Sensing	Friess et al. (2015)
Myanmar	Myeik /Taninthyari National Park	Natural—mixed spp.	not indicated	78.9	2016	Vegetation	Remote Sensing	Darmawan et al. (2016)

estimates reported. There are 6 featured Ramsar sites and 5 UNESCO Biosphere Reserves featured in this study where two share both classifications, and other protected mangroves in Southeast Asia.

In the featured mangrove Ramsar sites, the estimated blue carbon stocks ranged from 141 Mg C ha^{-1} (Sembilang National Park, Indonesia) to 1220 Mg C ha^{-1} (Tanjung Puting, Indonesia) (Manuri et al. 2011). Estimates for Sembilang National Park are focused on the vegetation only. Such is the case for Phuket-Phang Nga Bay in Thailand with a vegetation carbon stock of 166.3 Mg C ha^{-1}. For Koh Klang in Thailand, high carbon stock was reported for vegetation and sediment combined with 1030 Mg C ha^{-1}. Given these, those studies that focused on aboveground biomass are expected to have smaller values such as in the case of Johor Park, Malaysia. According to Rozainah et al. (2018) the carbon stock accumulated in Pulau Kukup and Tanjung Piai in Johor Park accounts for 20% only in vegetation and the rest are stored in the soil.

In terms of the Biosphere reserves, the estimated carbon stocks range from 77 Mg C ha^{-1} (in Kien Giang Province, Vietnam) to 1220 Mg C ha^{-1} (in Tanjung Puting, Indonesia). Tanjung Puting National Park in central Kalimantan, Indonesia was declared a UNESCO Biosphere Reserve in 1977 and was also recently designated as a Ramsar site as well in 2013. According to Murdiyarso et al. (2009), this site contains huge sediment carbon stock with 1060 Mg C ha^{-1} or 87% of the total estimate. This is highly attributable to its thick sediment (< 3 m) and rich carbon content. On the other hand, Wilson (2010) reported a very small stock for Kien Giang province in Vietnam with 76.9 Mg C ha^{-1} that accounts for both vegetation and sediment. Vegetation of this site was described to be young in age and sparse in cover. This therefore implies that for blue carbon stocks to increase, sustained site protection and assisted natural regeneration to improve the vegetation cover is needed. Such is the case for the dense mangrove forest of Mui Ca Mau National Park (both a Ramsar and a biosphere reserve). Carbon stock of this sites was estimated to 770 Mg C ha^{-1}. In view of Palawan, Philippines (which was enlisted to be a Man and Biosphere Reserve in 1990), blue carbon stocks was estimated to about 529.85 Mg C ha^{-1} from both vegetation and sediment pools (Abino et al. 2014).

In view of other protected areas, the natural mangrove forest of Peam Krasaop Wildlife Sanctuary in Koh Kong Province, Cambodia has about 953.9 Mg C ha¯1. As much as 90% this is shared by the sediment pool. For Pulau Ubin and Sungei Buloh in Singapore, the combined vegetation and sediment blue carbon stock ranged only from 163.7 to 249.3 Mg C ha^{-1}. Vegetation shared a larger portion of this stocks with about 75%.

Research on mangrove blue carbon has been growing and increasing over the past two decades. There are about 32 Ramsar sites and around 15 UNESCO Biosphere reserves which involve several mangrove forests in Southeast Asia. However, only a few of these sites have information about their blue carbon stocks. Further, updated information about their management status is also limited. The likely variations in values reported is also highly reflective of the assessment methods used. While some studies roughly estimated carbon stocks through remote sensing, many blue carbon studies followed several protocols that employed field measurements such as the one developed by Kauffman and Donato (2012). For instance, Darmawan et al. (2019) observed that carbon stock estimation using only remote sensing approach (through ALOS-PALSAR) have generated larger estimates than those obtained from actual field measurements. This therefore implies the need for combining or harmonizing both methods to achieve more accurate assessment. Furthermore, carbon stock estimates are likely to vary in terms of: (i) sampling designs; (ii) allometric equations used; (iii) laboratory analysis methods for sediment carbon stock determination; to name a few. This put forward the need to come up with a standardize method for mangrove blue carbon stock assessment like the one proposed by Kauffman and Donato (2012).

19.4 Challenges in Mangrove Blue Carbon Management

19.4.1 Coastal Development and Mangrove Conversion

Over the past five decades, the drastic decline in mangrove cover can be largely attributed to land use conversion. Replacing mangroves with other land uses spells loss in biodiversity and decrease in carbon emissions (Pendleton et al. 2012). One example is a town in Pahang, Malaysia where most of the mangroves have been cleared for coastal development. Saad et al. (1999) emphasized that such an event led to biodiversity loss and habitat degradation, displaced mangrove-dependent livelihoods, and diminished ecosystem services such as carbon sequestration since both vegetation and sediments were significantly disturbed. Same threats may likely face the conservation areas such as those under UNSESCO biosphere reserves and Ramsar sites. In Singapore, Ng and Low (1994) asserted that the indirect drivers of land use conversion should be clearly understood. These include the increasing coastal population growth. Loss in mangrove cover is associated with the expansion of housing, roads and other infrastructures towards the coastal areas (Fig. 19.3). Likewise, the widespread aquaculture contributed to massive mangrove loss in the Philippines and Vietnam (Primavera 2005b; Orchard et al. 2015). The increasing demand for shrimp, fish and crabs by the growing population, and the high demand for export of these products are considered to be foundational drivers of this threat. Further, mangrove conversion to rice fields contribute to mangrove decline in some provinces of Vietnam and Myanmar (Gevaña et al. 2018b).

19.4.2 Limited Incentive-Based Schemes for Mangrove Conservation

The usual response of imposing stricter policies to place mangroves under strict protection without any provision for alternative income sources has detrimental impacts on the poor coastal community's well-being. This discourages participation

Fig. 19.3 Road construction damaging natural mangrove areas. (Photo credit: Dixon Gevaña)

in mangrove conservation. According to Vanderklift et al. (2019), the conservation of coastal ecosystems is more likely to bear fruit with a good stakeholder engagement on various project and policy-making endeavors. A major challenge in blue carbon management is the generation of incentives for community-based mangrove conservation. One good opportunity is through the Payments for Ecosystem Services (PES) schemes. This aims to provide options of replacing resource-extractive livelihoods sources such as aquaculture and commercial firewood harvesting, while supplementing those that are for domestic and subsistent purposes only. According to Friess and Thompson (2016), mangrove PES is still in its early stage and is recently being adopted by several countries like Thailand. Similarly, several mangrove forests in Vietnam such as in Xuan Thuy National Park (Ramsar Site), Kien Giang and Nghe An had adopted PES programs in mangroves which are described to be needing continuous commitment both from government and local communities in order to sustain the income benefits (Hawkins and Cuong 2015). In Malaysia, Aziz et al. (2016) asserted that PES such as REDD+ projects would not entirely cease tree cuttings in mangroves but may boost ecological services by providing production and conservation stability. It was argued that many people in the rural coasts remain dependent on mangroves for firewood and resource-extractive livelihoods. Such a challenge needs to be sufficiently understood and responded to in crating alternative income sources through PES and other incentive-based conservation programs.

19.4.3 Unclear Tenure and Property Rights

It is always important to answer the question: *who owns the mangroves?* Low commitment in mangrove conservation is often reflective of tenurial and resource management issues. The overlapping resource rights and mandates among local communities, government, private sectors, and other stakeholders are reflective of unclear policies and institutional arrangements. It is therefore difficult to determine the authority and accountability over all the planning and decisions towards mangrove resources. Knox et al. (2011) stressed that it is also critical to establish resource ownership. This could help correctly attribute both the benefits and responsibilities of mangrove conservation efforts. This is also a major consideration in designing forest carbon projects where carbon rights should be clearly determined. Further, Larson et al. (2013) pointed that sustainable realization of benefits will only happen if there is a secured land rights. Conversely, deforestation and land grabbing will continue persist with tenure insecurity particularly against the poor local communities who are highly dependent on forest resources for a very long time (Bolin et al. 2013). This often results to. resource use and management conflicts.

19.4.4 Low Survival Rates in Mangrove Rehabilitation

Revegetating degraded and deforested mangrove areas has received much support and interest over the past two decades. A major challenge in most of mangrove reforestation projects is the long-term survival rates and revival of critical ecological functions. In many cases, the problem of low planting survival rate is associated with poor site-species suitability matching. In the Philippines, Primavera and Esteban (2008) reported a very low survival rate only about 20% for selected reforestation projects. Such undesirable result was due to common practice of monoculture planting particularly on zones where they are not suitable. Further, the problem also transpires from the lack or limited understanding of the degraded sites in terms of biophysical (eg. sediment quality, plant zonation, hydrology and biodiversity) and socio-economic factors that contribute to the further degradation and mismanagement (Camacho et al. 2020). Limited capacities in doing right nursery and planting work likewise add to unsuccessful mangrove rehabilitation. Designing restoration plans should therefore be responsive to future climate projections and their implied changes in coastal hydrology. This is likely to be a serious challenge for most mangrove planners and managers since technical knowledge on climate-proofing mangrove management plans to respond to sea level rise, storm surge projections, water salinity, to name a few, is always limited (Fig. 19.4).

19.5 Way Forward for Blue Carbon Conservation

Developing strategies for mangrove blue carbon conservation requires a good understanding of the drivers, pressures, state, issues and responses (DPSIR) to mangrove deforestation and degradation (Fig. 19.5). In general, coastal population growth and poor coastal governance contribute largely to the overexploitation of mangrove resources and their unabated conversion to other land uses such as aquaculture ponds, farms, settlement areas, roads and industries, to name a few.

Fig. 19.4 A planted site of *Ceriops spp* on unstable sediment is experiencing a low survival rate. (Photo credit: Dixon Gevaña)

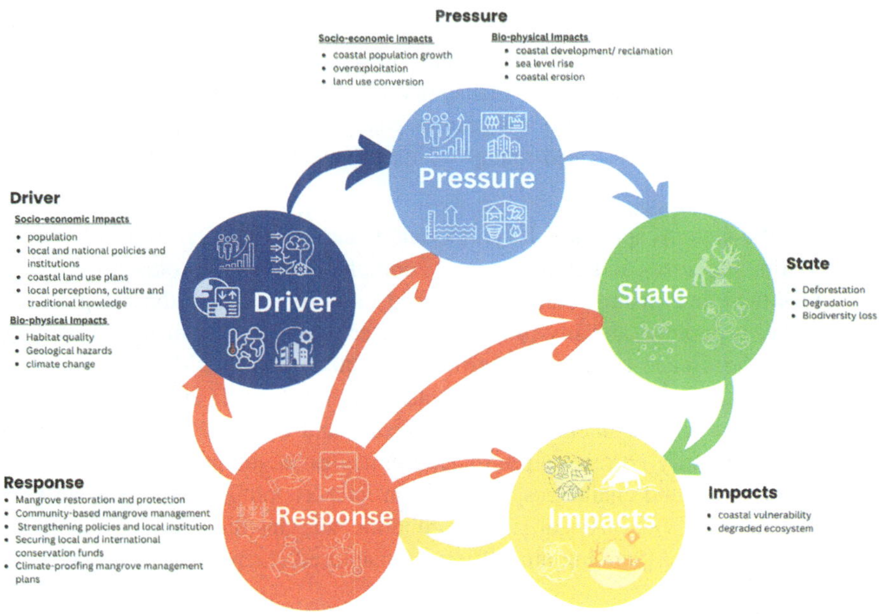

Fig. 19.5 Driyers, pressures, state, impacts, and responses to mangrove deforestation and degradation

Mangrove loss is likely to be intensified by the future impacts of sea level rise (Macintosh et al. 2023). Degraded coastal ecosystem also worsens the vulnerability of coastal communities and biodiversity (including their ecosystem services) to climate hazards. In view of the present efforts, many countries in Southeast Asia pursue mangrove restoration and protection programs through community-based approaches. There are also efforts to strengthen policies and local institutions to enhance mangrove protection. International collaborations such as the *Global Mangrove Alliance* are also in place. This alliance aims to brings together technical experts, civil society organizations, governments, local communities, businesses, funding agencies and foundations to accelerate a comprehensive, coordinated, global approach to mangrove conservation and restoration at relevant scales.

Making mangrove conservation happen from the international to local levels remains a big challenge. It necessitates: (i) development of integrated mangrove management plan; (ii) promotion of collaborative mangrove governance; and (iii) forging regional mangrove alliances.

19.5.1 Integrated Mangrove Management Plan

Developing a specific plan for mangrove management is ideal to sufficiently capture and communicate the status, problems and appropriate strategies for conservation. This plan should complement or add up to the existing broader conservation plans for biosphere reserves, RAMSAR sites and other protected areas. By doing these, strategic priorities for mangrove management (eg. development of blue carbon projects, ecotourism, to name a few) can be easily identified and pursued (Gevaña et al. 2022). A good integrated mangrove management plan should adhere to the following principles:

(i) Holistic systems approach. It is important to capture the interactions of various factors, processes and components, as complex social and ecological systems. This therefore ensures that concerns and interests spanning the bio-physical, social, economic, and institutional characteristics not only in mangroves but with other linked ecosystems upstream are taken into consideration and integrated into the strategy formulation and planning (Gevaña et al. 2021).

(ii) Science-based and interdisciplinary. Factual and evidence-based planning is vital to identify correct and sufficient strategies in solving past, present and future issues and challenges. Through an interdisciplinary approach, all relevant aspects (bio-physical, socio-economic and institutional) of mangrove assessment and planning are sufficiently captured (Liang et al. 2017; Gevaña et al. 2021).

(iii) Inclusiveness and participatory. Participatory epistemology is necessary to ensure the sustainability and inclusiveness of an integrated mangrove management and development effort (Gevaña et al. 2018b). Recognizing that mangroves host several groups of major stakeholder, their participation is imperative in resolving conflicting interests that may affect the efficient, effective and sustainable management and utilization of resources.

(iv) Needs and demand responsive. Present and future needs and demand of key stakeholders are of paramount focus of the plan. Responding to these will help boost commitment and participation towards mangrove conservation (Liang et al. 2017).

(v) Culture and gender sensitive. Needs and interests on mangrove resources are always driven by culture, traditional knowledges, gender roles and expectations. The Plan and future should therefore aim to strengthen and promote: cultural/customary rights; tenurial rights; and gender sensitivity and development (Camacho et al. 2020).

19.5.2 Collaborative Mangrove Management

Mangrove issues are complex, and they are driven by various interests. These issues must be addressed on the principle multi-stakeholderism and collaboration. Gevaña et al. (2021) proposes that collaborative management is one of the best approach to mangrove management. This intimately involves the local communities, local government and other major stakeholders in collectively identifying mangrove issues and craft policies and plans to address them. There are likely positive impacts of adopting this, such as: (i) strengthening of local community's sense of ownership over mangrove that have overlapping tenurial claims, organisation and responsibility over mangrove management; (ii) gender empowerment; (iii) better constituencies for the local government; (iv) active participation of non-government stakeholders, and (iv) creation of alternative livelihoods that are related to mangrove conservation; to name a few (Larson et al. 2012; Liang et al. 2017; Camacho et al. 2020).

19.5.3 Regional Mangrove Alliance and Road Map

Forming mangrove alliances could be helpful and strategic to identify conservation targets and solicit commitments among countries to promote blue carbon management. Most of the alliances are formed through international conferences and workshops which gather mangrove scientists, government agencies and non-government organizations who shared passion in mangrove conservation. Over the past decade, a number of regional mangrove meetings were organized. These include: *Regional Symposium on Mangrove Ecosystem Management in Southeast Asia* (February 2013, Surabaya Indonesia); *International Conference on Mangroves of Asia-Pacific in View of Climate Change* (November 2014, Kuala Lumpur Malaysia); *ITTO's* Conference *on Sustainable* Mangrove *Ecosystems* (April 2017, Bali Indonesia); *ASEAN Mangrove Congress* (December 2012 and September 2017, Manila Philippines); *Fifth International Mangrove Macrobenthos and Management Meeting* (July 2019, Singapore), to name a few. These initiatives provided a venue for sharing good practices in mangrove management, as well policy statements and recommendations for the ASEAN and respective governments to act in view of the most important issues concerning mangroves today and in the future. A critical step for the ASEAN is to respond by coming up with its roadmap for mangroves. This roadmap should outline the targets and commitment towards the next decades on: (i) mangrove restoration in view of the Global Mangrove Alliance goal of 20% increase in mangrove cover; (ii) science-based rehabilitation responsive of the future sea level rise projections; (iii) blue carbon investments; (iv) production and trade of mangrove commodities; and (v) institutionalization and harmonization of mangrove policies among member states.

19.6 Summary and Recommendations

Mangrove conservation has received greater attention over the past decades due to their critical ecosystem functions such as carbon sequestration, coastal protection from storm surge and tsunami, and supporting local livelihoods. The UN-led mangrove conservation programs in Southeast Asia have demonstrated a good case for reversing coastal deforestation trend. It also offers diverse opportunities for establishing blue carbon programs in the future. The huge amount of carbon stocks that were observed in several biosphere reserves, Ramsar sites and protected areas imply the need for serious conservation and expansion of conservation efforts towards areas where degradation remains. Foundational to pursuing this is the good understanding of the key issues and drivers of mangrove loss. These largely include those that are related to poor land use planning, limited incentives from mangrove conservation, tenurial insecurity, and lack of science-based/ecological considerations in rehabilitation. Some of the likely strategies to overcome these challenges are: (i) crafting of local integrated mangrove management plans; ii) adopting collaborative mangrove governance approach in all conservation and livelihood projects; and iii) forge regional alliances and mangrove road map for the ASEAN towards effectively pursuing international collaborations on blue carbon programs.

Acknowledgement This research is supported by the Ministry of Education, Singapore, under its Social Science Research Council 'Climate Governance of Nature-based Carbon Sinks in Southeast Asia' (MOE2021-SSRTG-021). Any opinions, findings and conclusions or recommendations expressed in this material are those of the author(s) and do not reflect the views of the Ministry of Education, Singapore.

References

Abino A, Castillo J, Lee Y (2014) Species diversity, Biomass, and carbon stock assessments of a natural mangrove forest in Palawan, Philippines. Pak J Bot 46:1955–1962

Adedeji O, Okocha R, Olatoye O (2014) Global climate change. J Geosci Environ Prot 02:114–122. https://doi.org/10.4236/gep.2014.22016

Alongi DM (2012) Carbon sequestration in mangrove forests. Carbon Manage 3(3):313–322. https://doi.org/10.4155/cmt.12.20

Aziz AA, Thomas S, Dargusch P, Phinn S (2016) Assessing the potential of REDD+ in a production mangrove forest in Malaysia using stakeholder analysis and ecosystem services mapping. Mar Policy 74:6–17. https://doi.org/10.1016/j.marpol.2016.09.013

Barrientos K, Apolonio J (2018) Species diversity and soil carbon sequestration potential of mangrove species at Katunggan It Ibajay (KII), Aklan, Philippines. Prism, p 22

Bolin A, Lawrence L, Leggett M (2013) Land tenure and fast–tracking REDD+: time to reframe the debate? Global Canopy Programme (GC)

Bunting P, Rosenqvist A, Lucas R, Rebelo L-M, Hilarides L, Thomas N, Hardy A, Itoh T, Shimada M, Finlayson CM (2018) The global Mangrove watch – a new 2010 global baseline of mangrove extent. Remote Sens 10(10):1669. https://doi.org/10.3390/rs1010669

Camacho L, Gevaña D, Sabino L, Ruzol C, Garcia J, Camacho A, OO T, Maung A, Saxena K, Liang L, You E, Takeuchi K (2020) Sustainable mangrove rehabilitation: lessons and insights from community-based management in The Philippines and Myanmar. APN Sci Bull 10(1): 18–25. https://doi.org/10.30852/sb.2020.983

Chowdhury RR, Uchida E, Chen L, Osorio V, Yoder L (2017) Anthropogenic drivers of mangrove loss: geographic patterns and implications for livelihoods. Mangrove ecosystems: a global biogeographic perspective, 275–300. doi:https://doi.org/10.1007/978-3-319-62206-4_9

Darmawan S, Takeuchi W, Nakazono E, Parwati E, Dien VT, Oo KS, Wikantika K, Sari DK (2016) Characterization and spatial distribution of mangrove forest types based on ALOS-PALSAR mosaic 25m-resolution in Southeast Asia. IOP Conf Ser: Earth Environ Sci 37:012035. https://doi.org/10.1088/1755-1315/37/1/012035

Darmawan S, Sari D, Takeuchi W, Wikantika K, Hernawati R (2019) Development of aboveground mangrove forests' biomass dataset for Southeast Asia based on ALOS- PALSAR 25-m mosaic development of aboveground mangrove forests' biomass dataset for Southeast Asia based on ALOS-PALSAR 25-m mosaic. J Appl Remote Sens 13:44519. https://doi.org/10.1117/1.JRS.13.044519

Dung LV, Tue NT, Nhuan MT, Omori K (2016) Carbon storage in a restored mangrove forest in Can Gio Mangrove Forest Park, Mekong Delta, Vietnam. For Ecol Manage 380:31–40. https://doi.org/10.1016/j.foreco.2016.08.032

Elwin A, Bukoski J, Jintana V, Robinson E, Clark J (2019) Preservation and recovery of mangrove ecosystem carbon stocks in abandoned shrimp ponds. Sci Rep 9:18275. https://doi.org/10.1038/s41598-019-54893-6

FAO (2005) Global forest resources assessment: thematic study on mangroves

FAO (2015) Global forest resources assessment 2015: Desk reference. Food and Agriculture Organization, Rome, p 64

FAO (2020) Global forest resources assessment 2020: main report. Rome. https://doi.org/10.4060/ca9825en

Fisher R (2014) Lessons learned from community forestry in Asia and their relevance for REDD? USAID-supported Forest Carbon, Markets and Communities (FCMC) Program, Washington DC, p 45

Friess DA, Thompson BS (2016) Mangrove Payments for Ecosystem Services (PES): a viable funding mechanism for disaster risk reduction? In: Renaud FG, Sudmeier-Rieux K, Estrella M, Nehren U (eds) Ecoysystem-based disaster risk reduction and adaptatation in practice, Advances in natural and technological hazards research; vol. 42. Springer, pp 75–98. https://doi.org/10.1007/978-3-319-43633-3_4

Friess D, Richards R, Phang H (2015) Mangrove forests store high densities of carbon across the tropical urban landscape of Singapore. Urban Ecosyst 19(2):795–810. https://doi.org/10.1007/s11252-015-0511-3

Friess DA, Rogers K, Lovelock CE, Krauss KW, Hamilton SE, Lee SY, Lucas R, Primavera J, Anusha R, Shi S (2019) The state of the world's mangrove forests: Past, present, and future. Annu Rev Environ Res 2019(44):89–115

Garnett S, Sayer J, Du Toit J (2007) Improving the effectiveness of interventions to balance conservation and development: a conceptual framework. Ecol Soc 12:2

Gevaña D, Camacho L, Pulhin J (2018a) Conserving mangroves for their blue carbon: insights and prospects for community-based mangrove management in Southeast Asia. In: Makowski C, Finkl C (eds) Threats to mangrove forests, Coastal Research Library, vol 25. Springer Nature, pp 579–588

Gevaña D, Pulhin J, Tapia M (2018b) Fostering climate change mitigation through a community-based approach: carbon stock potential of community-managed mangroves in The Philippines. In: Krishnamurthy J, Srinivasalu G (eds) Coastal management: global challenges and innovations. Elsevier, pp 271–282

Gevaña D, Garcia J, Ruzol C, Malabayabas F, Grefalda L, O'Brien E, Santos E, Camacho L (2021) Climate change resiliency through mangrove conservation: the case of alitas farmers of Infanta, Philippines. In: Nishi et al (eds) Fostering transformative change for sustainability in the context of Socio-ecological Production Landscapes and Seascapes (SEPLS). Springer Nature, pp 195–214. https://doi.org/10.1007/978-981-33-6761-6_12

Gevaña D, Villanueva C, Garcia J, Camacho L (2022) Mangroves sustaining biodiversity, local livelihoods, blue carbon, and local resilience in Verde Island passage in Luzon, Philippines. In: Das SC, Pullaiah A, E.C. (eds) Mangroves: biodiversity, livelihoods and conservation. Springer Nature, Singapore. https://doi.org/10.1007/978-981-19-0519-3_17

Giri C, Ochieng E, Tieszen L, Zhu Z, Singh A, Loveland T, Masek J, Duke N (2011) Status and distribution of mangrove forests of the world using earth observation satellite data. Glob Ecol Biogeogr 20:154–159

Goldberg L, Lagomasino D, Thomas N, Fatoyinbo L (2020) Global declines in human-driven mangrove loss. Glob Chang Biol 26:5844–5855. https://doi.org/10.1111/gcb.15275

Hamilton SE, Casey D (2016) Creation of a high spatio-temporal resolution global database of continuous mangrove forest cover for the 21st century (CGMFC-21). Glob Ecol Biogeogr 25: 729–738

Hawkins S, Cuong C (2015) Roots in the water: legal frameworks for mangrove PES in Vietnam

Heshmati H (2020) Impact of climate change on life. IntechOpen. https://doi.org/10.5772/intechopen.94538

Jentoft S (2005) Fisheries co-management as empowerment. Mar Policy 29:1–7. https://doi.org/10.1016/j.marpol.2004.01.003

Kathiresan K, Bingham B (2001) Biology of mangroves and mangrove ecosystems. Adv Mar Biol 40:81–251. https://doi.org/10.1016/S0065-2881(01)40003-4

Kauffman JB, Donato D (2012) Protocols for the measurement, monitoring and reporting of structure, biomass, and carbon stocks in mangrove forests. Working Paper 86. Center for International Forest Research, Bogor

Knox A, Caron C, Miner J, Goldstein A (2011) Land tenure and payment for environmental services: challenges and opportunities for REDD+. Land Tenure J 2

Larson A, Brockhaus M, Sunderlin W (2012) Tenure matters in REDD+: lessons from the field. In: Angelsen A, Brockhaus M, Sunderlin WD, Verchot L (eds) Analysing REDD+: Challenges and choices [Japanese]. Center for International Forestry Research (CIFOR), Bogor, pp 153–176

Larson A, Brockhaus M, Sunderlin W, Duchelle A, Babon A, Dokken T, Pham T, Resosudarmo I, Selaya G, Awono A, Huynh T (2013) Land tenure and REDD+: the good, the bad and the ugly. Glob Environ Chang 23(3):678–689. https://doi.org/10.1016/j.gloenvcha.2013.02.014

Liang L, Yiu E, Camacho D, Gevaña D, Oo T, Sein C, Chen B, Nakama Y, Saxena K, Takeuchi K, Nagata A (2017) Coastal forest management in the face of global change: experience of four Asian countries. APN Sci Bull 7(1):8. https://doi.org/10.30852/sb.2017.70

Macintosh D, Salmo S, Gevaña D, Sidik F, Dharmawan I, Nightingale M, Valderrabano M (2023) Mangroves of the wester coral triangle. IUCN red list of ecosystem assessments. International Union for Conservation of Nature IUCN, Switzerland. https://doi.org/10.32942/X21K5P

Manuri S, Putra C, Imansyah T, Purbopuspito J, Warren M (2011) C-stock assessment of mangrove ecosystem at sembilang national park, South Sumatra. https://doi.org/10.13140/2.1.3600.8965

McKee K (1996) Mangrove ecosystems: definitions, distribution, zonation, forest structure, trophic structure and ecological significance. Mangrove ecology: a manual for a field course mangrove. Smithsonian Institution, Washington

Mcleod E, Salm R (2006) Managing mangroves for resistance to climate change. IUCN, Gland

Murdiyarso D, Donato D, Kauffman J, Kurnianto S, Stidham M, Kanninen M (2009) Carbon storage in mangrove and peatland ecosystems. A preliminary account from plots in Indonesia, Working paper 48. Center for International Forestry Research, Bogor Banat, 35 p

Nagelkerken I, Blaber S, Bouillon S, Green P, Haywood M, Kirton L, Meynecke J, Pawlik J, Penrose H, Sasekumar A, Somerfield P (2008) The habitat function of mangroves for terrestrial and marine fauna: A review. Aquat Bot 89:155–185. https://doi.org/10.1016/j.aquabot.2007.12.007

Ng P, Low J (1994) Status of mangroves in Singapore: conservation beyond the year 2000. National University of Singapore, Kent Ridge

NOAA (2020) State of the climate: global climate report for annual 2020. National Centers for Environmental Information. https://www.ncdc.noaa.gov/sotc/global/202013

Oo NW (2002) Present state and problems of mangrove management in Myanmar. Trees 16: 218–223

Orchard SE, Stringer LC, Quinn CH (2015) Impacts of aquaculture on social networks in the mangrove systems of northern Vietnam. Ocean Coast Manag 114:1–10. https://doi.org/10.1016/j.ocecoaman.2015.05.019

Pendleton L, Donato D, Murray B, Crooks S, Jenkins W, Sifleet S, Craft C, Fourqurean J, Kauffman J, Marba N, Megonigal P, Pidgeon E, Herr D, Gordon D, Baldera A (2012) Estimating global "Blue Carbon" emissions from conversion and degradation of vegetated coastal ecosystems. PLoS One 7:e43542. https://doi.org/10.1371/journal.pone.0043542

Primavera JH (2005a) Mangroves, fishponds, and the quest for sustainability. Science 310:57–59

Primavera J (2005b) Mangroves and aquaculture in Southeast Asia. Report of the regional technical consultation for the development of code of practice for responsible aquaculture in mangrove ecosystems

Primavera JH, Esteban JMA (2008) A review of mangrove rehabilitation in The Philippines: successes, failures and future prospects. Wetl Ecol Manag 16:345–358

Pulhin J, Gevaña D, Pulhin F (2017) Community-based mangrove management in the Philippines: experience and challenges in the context of changing climate. In: Participatory mangrove management in a changing climate, pp 247–262. https://doi.org/10.1007/978-4-431-56481-2_16

Richards DR, Friess DA (2016) Rates and drivers of mangrove deforestation in Southeast Asia 2000–2012. PNAS 113(2):344–349. www.pnas.org/cgi/doi/10.1073/pnas.1510272113

Rozainah MZ, Nazri MN, Sofawi AB, Hemati Z, Juliana WA (2018) Estimation of carbon pool in soil, above and below ground vegetation at different types of mangrove forests in Peninsular Malaysia. Mar Pollut Bull 137:237–245. https://doi.org/10.1016/j.marpolbul.2018.10.023

Ruzol C, Camacho A, Sabino L, Garcia J, Gevaña D, Camacho L (2020) A materialist-idealist divide? Policy and practice in participatory mangrove rehabilitation in The Philippines. Environ Sci Pol 112:394–404

Saad S, Husain ML, Yaacob R et al (1999) Sediment accretion and variability of sedimentological characteristics of a tropical estuarine mangrove: Kemaman, Terengganu, Malaysia. Mangrove Salt Marshes 3:51–58. https://doi.org/10.1023/A:1009936014043

Sharma S, MacKenzie RA, Tieng T, Soben K, Tulyasuwan N, Resanond A, Blate G, Litton CM (2019) The impacts of degradation, deforestation, and restoration on mangrove ecosystem carbon stocks across Cambodia. Sci Total Environ 135416. https://doi.org/10.1016/j.scitotenv.2019.135416

Spalding M, Kainuma M, Collins L (2010) World Atlas of Mangroves. Earthscan, London

Tan D, Wan A, Wan J, Maimon A (2012) Community structure and productivity of mangrove forests in two National Parks of West Malaysia. Malaysian For 75:165–176

Thomas N, Lucas R, Bunting P, Hardy A, Rosenqvist A, Simard M (2017) Distribution and drivers of global mangrove forest change, 1996-2010. PLoS One 12:e0179302. https://doi.org/10.1371/journal.pone.0179302

Tue N, Dung L, Nhuan M, Omori K (2014) Carbon storage of a tropical mangrove forest in Mui Ca Mau National Park, Vietnam. Catena 121:119–126

Vanderklift M, Marcos-Martinez R, Butler JA, Coleman M, Lawrence A, Prislan H, Steven A, Thomas S (2019) Constraints and opportunities for market-based finance for the restoration and protection of blue carbon ecosystems. Mar Policy 107:103429. https://doi.org/10.1016/j.marpol.2019.02.001

Wilson N (2010) Biomass and regeneration of mangrove vegetation in Kien Giang Province, Vietnam https://doi.org/10.13140/RG.2.2.12896.17921

Index

M. Clüsener-Godt et al. (eds.), *Blue Carbon Mangrove Ecosystems*, Blue Carbon Ecosystems for Sustainable Development 1,
https://doi.org/10.1007/978-3-031-69553-7

The manufacturer's authorised representative in the EU is Springer Nature Customer Service Centre GmbH, Europaplatz 3, 69115 Heidelberg, Germany. If you have any concerns regarding our products, please contact ProductSafety@springernature.com

Printed and bound by CPI Group (UK) Ltd, Croydon, CR0 4YY
28/09/2026
02224081-0007